This book was written to help students prepare for the "APES Exam".   Let's be honest, we all know APES is an acronym for AP Environmental Science. AP and Advanced Placement are registered trademarks of the College Entrance Examination Board. The College Board was in no way involved with the production of this book. They probably don't even know it exists, and if they did they definitely wouldn't endorse it! Why would they? They didn't have anything to do with its production. Remember?

I would like to thank my wonderful wife and kids for putting up with me through the writing process. I would also like to thank a long list of encouraging life mentors and friends such as; my parents, Wiley McCauley, Dr. George Damoff, Brian Kaestner, Carolyn Schofield, and countless others who have given their best attempt to make something useful out of me. Benjy

I would like to thank Nicki, Holice, and Joe. This is the varsity team! I would also like to thank my parents for twenty-three chromosomes apiece and for their unconditional support. I would also like to thank Dr. Horner for not kicking me out of the graduate program, since this book has consumed all of my "thesis-writing" time. Dan Patrick gets special thanks for editing this book and making it almost readable. Dave

ISBN 978-0-557-11922-6

Illustrations by Dave Holbert

Cover art by Olof Olsson and Cindy Bai

Gorilla on p. 124 by Ravin Ricord

THIS BOOK IS NOT A BLACKLINE MASTER. PLEASE DO NOT COPY! WE ARE POOR TEACHERS TOO!

APES

in a BOX

Benjy Wood and Dave Holbert

Greetings,

We begin this book as two AP Environmental Science teachers facing the annual questions that every APES teacher faces: how do we best review our students for the APES exam, and which review materials are best for our students? Every year we find that existing review materials are too elementary, conveniently too much like the AP exams that have been released, or are longer and more complex than the textbook! Recently we decided to stop asking the question of which review materials are best, and start writing our own. Many late nights, some broken computer equipment, and a partial ownership of a small coffee importer later, APES in a Box was born. The premise behind the book is simple. Wouldn't you like to get your hands on the notes of that curve-buster that sits in the front row? We wrote this book to look like that perfect set of notes. From cover to cover you will find complex topics simplified through easy to understand notes and original artwork. We hope that you will find the book to be concise, original, on target, and most of all useful in preparing for the APES exam.

Good Luck and Many Happy 5's,

Benjy and Dave

The Exam

Before you begin to cram for the exam, let's look the beast in the eye to see what you're really up against. Here we'll break down the multiple choice and free response portions of the exam, and begin to develop some test day strategies to help take your score to the next level. First let's deal with the evil badness. Oops, I mean the multiple choice.

Multiple Choice

When properly prepared for the exam, students often tell us that the most difficult portion of the test is the multiple choice (MC). This portion of the exam requires that the student be able to recall and apply very specific information. Knowing what to expect will go a long way toward alleviating test anxiety and preparing you to make the most of the multiple choice. This portion of the exam consists of 100 multiple choice questions that you will have 90 minutes to answer. There are several types of mc questions which we have included.

On the multiple choice portion of the exam time is your enemy! It will be very difficult to answer every question with less than a minute given per question. We recommend going through the exam twice. The first time, skip questions that you cannot answer quickly. Once you have gone through the exam and have answered all of the "easy ones", use the remaining time to go back and answer the remaining questions. If the clock is ticking down and all else fails, make sure that you bubble in answers to all questions. DO NOT LEAVE ANY QUESTIONS BLANK.

Multiple Choice Questions

The Question Set: There will be question sets that use the same five answer choices to answer multiple questions. Here is an example:

Use the following answer choices to answer questions 1-3 .

    a. Hydrothermal energy

    b. Wind generated electricity

    c. Photovoltaic cells

    d. Passive solar energy

    e. Biomass energy production

1. Refers to the use of organic matter to produce energy or electricity.

2. Collects energy from the sun, and converts it into electricity.

3. Directly uses the suns energy to heat water or air.

The Multiple-Multiple Choice: A lot of time can be wasted on the multiple-multiple choice by going back and forth between the roman numerals and the answer choices. The best way to tackle this question is to mark each Roman numeral as being true or false. When you get to the end, find the answer choice that corresponds with your true or false analysis. In the problem below I-III are true, and IV is false, therefore the correct answer choice is C. I, II, and III only.

4. Which of the following are problems associated with the use of monoculture crops in agriculture?

       I. Increased use of pesticides

      II. Increased use of fertilizers

     III. Increased use of irrigation

     IV. Decreased food production

a. I only

b. I and II only

c. I, II, and III only

d. I, II, and IV only

e. I, II, III, and IV

The Traditional Multiple Choice With Five Answer Choices

5. All of the following are problems associated with construction of dams **except:**

a. Regular downstream flooding.

b. Trapping of sediment behind the dam.

c. loss of habitat from the formation of a reservoir.

d. The high initial cost of building a dam.

e. The disruption of migratory routes for certain species of fish.

The Free Response

The free response questions are graded with a rubric which means that you will be awarded points based on your correct answers. Points are not deducted for incorrect answers. Because of this, you should attempt to answer every portion of the free response. Let's break down the free response.

You will have 90 minutes to answer the four free response questions, and you will not be allowed to use a calculator. Later, we will look at each type of question that is likely to show up on the free response portion. First, here are our top ten keys for success on the free response:

1. When you begin the test, quickly read all of the questions, and rank them in order of difficulty. Begin working on the "easiest" question, and work your way toward the more difficult ones. Since this is a timed test, you would rather spend your time working on the questions that you will most likely be able to complete successfully. If you do run out of time, you would rather run out of time on the most difficult question.

2. Answer the question in the order that it is given to you. If the question has parts A, B, C, D, and E, then your answer should also have parts A, B, C, D, and E. Go ahead and label each part of your answer and leave a space in between each one. This will make it easier for the grader to read your paper and award you with the points that you have earned. If your answer is written in one giant paragraph, it makes it much more difficult to grade.

3. Stay on task! While you are writing, ask yourself whether or not what you are writing is directly answering the question. If the answer is no, then you shouldn't be writing it. Stay away from flowery introductions, and don't restate the

question. ONLY ANSWER THE QUESTION. Points are awarded for the correct answers that are found on the grading rubric. It only wastes time to write extra stuff.

4. Pay very close attention. This one is important. You need to be specific in your answers. For example, if a question asked you to "describe an environmental impact of using fossil fuels for transportation" you might answer that using fossil fuels generates air pollution. This would be a true statement, but it would not earn any points. You need to be more specific. You should explain that burning fossil fuels generates CO_2 emissions, increasing greenhouse gases, or that burning fossil fuels generates SO_2, which can contribute to acid rain. Specific answer = points.

5. From time to time you will be asked to draw a graph. Make sure that you give your graph a descriptive title. Your title should explain what is happening in the experiment. Make sure to label your axes appropriately. Your label should include what is being measured and the units of measurement. For example, Time (seconds).

6. Write your answers in complete sentences. The grader cannot assume anything, and can only give you credit for what you write. The more descriptive you are, the more likely you are to earn points.

7. Show all work on calculations. If a free response asks you to calculate something, there is probably a point given for the correct answer and a point given for showing all of your work.

8. Remember that the free response is graded with a rubric, and you will be awarded points for the things that you write correctly. You will not be penalized

for incorrect answers. With that in mind, you should attempt every portion of each of the free response questions.

9. Lately there have been some pretty complicated schematic diagrams on the free response portion of the test. There are boxes with arrows pointing every which way, and they often confuse students to the point that they don't really attempt the question. If you see a complicated schematic diagram, try to analyze it as a generic system. Follow the arrows and ask yourself what is going into the system, what is happening in the system, and what is coming out of the system. This will usually help you identify the important parts of the diagram, and help you formulate your answer.

10. Make your grader happy! Imagine that your paper is going to be graded by a very tired grader with eye fatigue. Write your answer concisely and neatly. Make it as easy as possible for your grader to read your answers.

The Free Response Types: There are four questions on the free response portion of the exam, and they usually follow these three question formats.

1. The document based question (DBQ): The DBQ begins with an article to read. The article is usually a fictitious article from a newspaper, and is usually about half a page long. The article will outline a particular environmental problem. You will then be asked to answer a set of questions that are related to the article. The article is there to set up the problem. We would not recommend quoting the article in your answer. You are expected to come up with your own answers, and will not usually get points for answers that are taken directly from the reading.

2. The calculation question: There will most likely be one question that requires a great deal of calculation. Most of the points in this question will be awarded for correctly setting up the problems, and correctly answering them. If you are a math whiz, then this is your time to shine. If math freaks you out, do not skip this question. Check all of the parts of the question. Some of the parts may not be calculation based at all, and you may still be able to earn some points on the question. Sometimes you will be asked to draw a graph. If this is the case, then be sure to appropriately title the graph and label the axes.

3. The content questions: Two of the four questions generally require you to have some specific content knowledge. For example, you may be asked to describe a process such as wastewater treatment. Further in the question you may also have to know some advantages and disadvantages of the wastewater treatment process, some alternatives to traditional wastewater treatment, and maybe even describe a law that relates to wastewater treatment.

Sample Free Response

During the 1930s, biologists introduced pheasants onto an island in Washington State. Using the following data, on the island's pheasant population, answer the questions below.

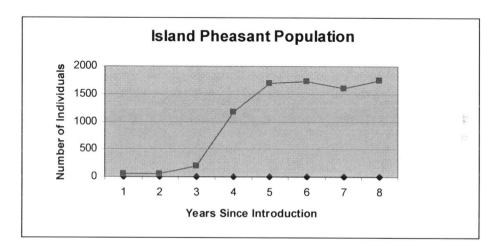

A. What is the approximate carrying capacity for the pheasant population?

B. Population density refers to the number of individuals living in a particular area. If the size of the island is 25 km², then what is the population density (per km²) during year seven? Please show all work.

C. Describe what the birth and death rates of the pheasant population must have been like during years 3-5. What other factors may have contributed to the rapid population growth seen during this period?

When preparing for the AP exam, you should be comfortable with the following information about the topics in this book, in your textbook, or in your APES class. For each topic you should know at least one or two examples of these types of information:

Environmental impacts
Economic impacts

Social and Cultural impacts
Legislation related to each topic

For example, you should be able to describe the environmental impacts of, the economic impacts of, and legislation related to pesticide use. Analyze each problem as if it were a generic system that you were trying to describe. Ask yourself what are the inputs (things going into the system), what is going on in the system, and what are the outputs (things leaving the system) of the system. These three things will make up the backbone of any free response question about an environmental problem.

Earth

In this first section of the book, we will take a look at the physical aspects of the planet. We will begin with geology, the biomes of the earth, and the way that matter and energy flow through the biosphere. We will then look at the way these things influence agriculture and land management.

Soil and Agriculture

Throughout most of our relatively short human history, we have survived as **hunter-gatherers**. Approximately 10,000-12,000 years ago, we began to practice early forms of agriculture. This was the beginning of the **Agricultural Revolution**. Humans began to increase food production through advanced cultivation techniques and selective breeding. Throughout most of the agricultural revolution, humans have practiced subsistence farming. This mainly involved family farms that existed for the purpose of feeding the family. Subsistence farmers may also sell, or barter, surplus crops. During the **Industrial Revolution** fossil fuels and machinery were applied in the pursuit of expanding world food production. These advances marked the beginning of the end for family farms. It became much more cost effective to produce food on large farms using these modern farming techniques. The **Green Revolution**, beginning in the 1940s, saw an even greater increase in food production with genetically altered crops, improved fertilizers, pesticides, and cultivation techniques. We are currently enjoying the successes of the Green Revolution, and trying to figure out how to deal with the negative impacts of it. There appears to be a resurgence of interest in more traditional farming techniques and organic farming. The challenge for the future will be to meet the increasing demand for food while practicing environmentally sustainable agriculture.

With the history lesson out of the way, let's take a quick look at what makes agriculture tick. Soil is often underappreciated. But is serves as the anchor and nutrient source for the plants that feed us, give us oxygen, and help to filter our land and water. Soil is composed of both inorganic and organic components. The particles that make up soil in order of decreasing size are: gravel, sand, silt, or clay. In the **Soil Triangle** below, see how the ratio of sand, silt, and clay determines the type of soil.

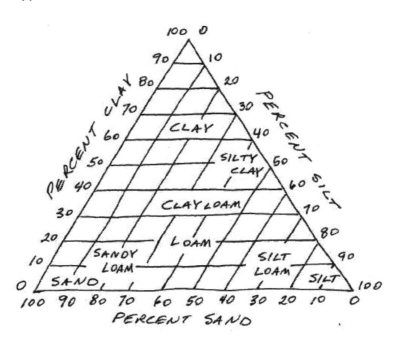

The organic portion of soil includes living organisms and humus. Humus is partially decomposed organic material. Some of the living organisms that can be found in soil include plants, nematodes, arthropods, fungi, bacteria, annelids, and gastropods. In the diagram below, you can see the different layers of soil that can generally be found.

SOIL HORIZONS

O-HORIZON
ORGANIC MATERIAL

A HORIZON - TOPSOIL

B HORIZON - SUBSOIL

C HORIZON - BEDROCK

Soil Degradation

Soil quality can be degraded by erosion, salinization, intensive agriculture, overgrazing, and development. Generally speaking, any activity that removes natural plant cover can result in soil degradation.

Soil erosion is simply the movement of soil from one place to another by wind or water. Erosion can be very costly to Farmer Brown as wind and water can remove precious topsoil from his farm. On the flipside, the soil is carried somewhere. If you happen to live downstream from Farmer Brown, his loss may be your gain if the rich topsoil happens to be **deposited** on your land! Be sure to read about ways to combat erosion in the agriculture section of this book. The main culprit in soil erosion is the removal of plant cover from the land. When plants are removed, the root systems are no longer in place to hold the soil together. Plants also play a crucial role as windbreaks, naturally reducing wind erosion. During the 1930s, drought, overgrazing, and over cultivation in the Southern Great Plains led to the historical period now referred to as the **Dust Bowl**. These conditions led to rapid erosion of topsoil which drastically decreased the productivity of the land and the ability of people to survive in the region.

Soil erosion can lead to major societal problems if it decreases crop yield. Haiti is currently undergoing a food shortage caused primarily by erosion. Years of deforestation have led to decreased plant cover, which has increased erosion. As the rate of erosion has increased, the productivity of Haiti's croplands has decreased. Some residents have resorted to eating a cake made of clay, salt, and lard in order to supplement their diets. Deforestation, overgrazing, and intensive cultivation reduce natural plant cover, and therefore contribute to soil erosion. Soil erosion is the main cause of soil degradation worldwide. Extreme soil erosion is a contributing factor in **desertification**. Desertification is simply the development of desert where there was not one before. Erosion contributes to this process by washing away precious topsoil, making it more difficult for plants to thrive. As fewer plants grow, more soil is washed away.

Soil erosion due to agriculture can be prevented by no-till farming, contour plowing, using shelterbelts, terracing, planting cover crops, and reforestation efforts.

No-till farming involves the use of special equipment to plant crop seeds directly into untilled soil. This practice drastically reduces the impact of soil erosion as there is very little soil that is disturbed. The process requires an initial investment in new agricultural equipment. No-till farming also requires greater application of herbicides to control weeds that would have otherwise been tilled into the soil.

Contour Plowing and planting involves planting crops perpendicular to the slope of the land. This allows each row of plants to serve as a barrier to erosion instead of an alley for erosion.

Shelterbelts or windbreaks are trees that are planted around the perimeter of cropland, or even in rows in between crops **(agroforestry)**. The trees help to block wind, which prevents wind erosion. The roots of the trees help hold soil in place and further help prevent erosion.

Terracing allows farmers to plant crops on steep slopes. Terraces are essentially shelves that are built perpendicular to the slope of the hill. The crops are grown on these shelves, which also serve to trap precious soil as it erodes down the slope.

Cover Crops can be planted during the offseason to help hold soil in place. This can include planting a crop that can be sold, and crops, like legumes, that return certain nutrients to the soil.

Soil Salinization is another cause of soil degradation. Continual irrigation of land can contribute to soil salinization. When a field is irrigated the water leaves dissolved salts on the surface of the soil. Rapid evaporation can also pull water, and its associated salts, from lower layers to the soil surface. Ultimately, the higher salt levels make it difficult for many plant species to thrive. Potential methods of dealing with salinization include the planting of salt tolerant plants, flushing the soil with water that has a low salt content, or letting the land rest while rainfall flushes out the excess salt. Salinization is a major contributor to crop loss in the United States.

Sometimes soil can be over-irrigated to the point that water in the soil prohibits proper gas exchange in the roots of plants. This condition is called **water-logging** and can result in crop loss.

Soil can be remediated by using organic fertilizers such as compost, **manure**, or **green manure**. This increases the organic material in the soil, decreases soil compaction, and increases the water holding capacity of the soil. All of these things will favor increased plant growth. **Inorganic fertilizers** are frequently used in soil remediation, but actually increase soil compaction, decrease the water holding capacity of soil, and generally only provide nitrogen, potassium, and phosphorus. Inorganic fertilizers are not ideal, but are inexpensive, easy to transport, and have drastically increased the world's food production.

Genetically modified organisms, or GMO's, have played a major role in the Green Revolution. GMO technology involves taking beneficial genes from a plant and inserting that beneficial gene into the genome of a crop species. This may include the insertion of genes for drought resistance, pest resistance, the need for less fertilizer, or even genes that produce extra nutritional value in the crop. The potential for GMO's is great as they have led to an increase in food production, particularly in areas that have struggled to produce enough food. They generally grow faster and have higher crop yields. Some suspect that GMO crops may have lower nutritional value, create new food allergens, and lead to the development of new plant toxins. GMO crops definitely lead to lower biodiversity in croplands.

Pest Control

Pesticides have been used for hundreds of years to increase crop yields. Pesticides include herbicides, insecticides, rodenticides, algicides, and fungicides. Using pesticides in agriculture began several thousand years ago by dusting plants with sulfur. This practice graduated to the use of chemicals like arsenic, lead, and mercury. You can probably guess why this wasn't such a good idea! Eventually we started looking to nature for plants that did not seem to have many pest problems and from these plants we extracted chemicals to apply to croplands. During the 1900s we learned how to make synthetic pesticides such as DDT. These pesticides are extremely effective, but are not without problems. It was discovered that some of these pesticides could take part in a process called biomagnification. **Biomagnification** occurs when a chemical enters living organisms such as algae, and then concentrates in the bodies of other organisms as the chemical travels up the food chain. DDT is one of these chemicals, and was shown to biomagnify until it caused reproductive problems in large predatory birds at the top of the food chain.

Some problems with pesticides include the development of pesticide resistance, high cost of application, and potential harmful effects on wildlife and humans when pesticides enter the ecosystem. **Pesticide resistance** occurs when pesticides are applied, susceptible organisms are killed, and non-susceptible organisms are left behind. These "pesticide resistant" organisms survive to pass on their genes to the next generation. Over time, the pesticide becomes useless as the pest population evolves resistance. This process is often referred to as the "pesticide treadmill" as the development of pesticide resistance requires increased application of pesticides, which in turn leads to more resistance!

Biological Pest Control and Integrated Pest Management

There are several pest control methods that do not require the use of chemical pesticides. Some examples include the introduction of pest predators, using pest pheromones to attract the pests to traps, using pest hormones to disrupt the pest's life cycle, or using crop rotation to combat specialist pest species. Perhaps one of the more devious biological pest management strategies involves the release of sterile male insects into the general pest population. The females mate with the sterilized males, assume that they have mated successfully, and then miss the breeding season. **Integrated Pest Management (IPM)** involves multiple strategies to reduce the use of pesticides. With an IPM program the goal is not necessarily to eradicate a pest species, but to reduce the pest damage to an acceptable level. With IPM the farmer uses cultivation techniques that disrupt pests, biological pest control methods, and if necessary they may use small amounts of narrow-spectrum pesticides. IPM provides great hope for an ecologically sustainable pest management option. Disadvantages of an IPM program include a high initial investment and education for farmers. IPM also takes a longer amount of time to be effective than traditional practices.

Meat production

Globally the consumption of animal products is on the rise. This includes an increase in the consumption of meat, eggs, and milk. This increased production has required an increase in cropland in order to supply the grain to feed the livestock. A large percentage of meat is produced on large factory farms. On these farms, chickens, pigs, or cattle are often crammed into small spaces where they are fed grain, which requires a significant amount of cropland to grow, and held until they are ready for processing. The main advantage of "factory farming" is the increase in yield and decrease in cost of the animal products that we consume. On a positive note, industrialized meat production has provided high quality protein at reasonable prices to millions of people.

Disadvantages of Factory Farms and Feedlots

Concentrating so many animals in one place can lead to contamination of groundwater when runoff, containing fecal coliform bacteria, reaches the water.

Animal waste can be a great source of fertilizer, but contributes to cultural eutrophication when runoff from feedlots enters lakes and streams.

Some techniques that are used to increase yield of meat, such as using antibiotics and hormones, pose potential health risks to consumers.

Some contend that animals, on industrialized farms, are treated poorly prior to processing.

Meat production also poses a resource problem. When we consider the pyramid of energy, and the fact that approximately 90% of energy is lost as heat as you move from one trophic level to the next, we must consider that it is more efficient to eat plant products than animal products. It takes a tremendous amount of grain to produce a small amount of meat. Beef is the most resource-intensive meat while chicken and fish use significantly less grain to produce the same amount of meat.

The Fine Print

The **Federal Insecticide, Fungicide, and Rodenticide Act (FIFRA)** regulates the distribution, sale, and use of pesticides. The act also requires large pesticide users to register their purchase.

The **Federal Food, Drug, and Cosmetic Act** give the EPA the power to establish safe levels for pesticides used in food and animal feed.

Fisheries and Aquaculture

Fish and seafood consumption has risen dramatically since the 1950s. As larger percentages of the world's population have developed a taste for seafood, the commercial fishing industry has continually developed new ways to increase its harvest. This has negatively impacted the ocean's fisheries. When a common public resource is exploited it is sometimes referred to as **The Tragedy of the Commons**.

Commercial Fishing Techniques

Trawling involves dragging a net across the ocean floor in an effort to catch bottom-dwelling species such as shrimp, flounder, and cod. Unfortunately the net has the potential to destroy just about anything on the ocean floor, including coral reefs, as it is pulled behind the boat. The main problem with trawling is by-catch.

By-catch refers to the non-target species that are caught accidentally in the pursuit of the target species. For example, many pounds of crab, fish, and other species are caught, killed, and thrown overboard for every pound of shrimp that reaches the consumer. New technology has produced nets that significantly reduce by-catch by allowing juvenile fish to pass through the nets unharmed.

Long line fishing is used to target fish like tuna. The fisherman drags a "long line" with several thousand baited hooks attached to it. Longlines can accidentally snag turtles, and seabirds, as the line is pulled through the ocean.

Driftnetting is a practice that involves setting large drift nets that can be several miles long. Fish swim into the net and become entangled. However, drift nets can also snag turtles and seabirds.

In **purse-seine fishing**, the fisherman use a large net to scoop up large schools of fish, such as yellow fin tuna, as they feed near the surface. Sometimes a spotter airplane is used to locate the fish for the boat. Dolphins will often swim above schools of tuna, and are inadvertently trapped by the net.

Most of the modern commercial fishing methods can be quite destructive, but the fact remains that we need food and the ocean provides a great deal of it. **Aquaculture**, or fish farming, is one possible way to meet the demand for seafood without many of the negative impacts of the commercial fishing process. Aquaculture involves raising fish, or other seafood, in nets that are either floating in open water, enclosing areas along a coast, or in inland ponds.

Advantages of Aquaculture

Fish farming reduces the pressure that is placed on fisheries by traditional commercial fishing methods.

Aquaculture does not use the vast quantity of fossil fuels that traditional fishing methods use.

Aquaculture produces a great deal of seafood in a small amount of space resulting in higher profits.

The Downside of Aquaculture

When high volumes of fish are concentrated in one area, they are more susceptible to disease.

Fish farms produce a high volume of waste, and water pollution, due to the high density of organisms confined to a small space.

Fish farms are often built in coastal areas which can disrupt mangrove forests and important breeding grounds for many species. The destruction of mangroves also makes the coastline vulnerable to hurricanes, tsunamis, and storm surges.

Fish farms often use grain to feed their fish which places further stress on the agricultural system.

The Fine Print

The International Convention for the Regulation of Whaling established the International Whaling Commission. The IWC is an international organization that regulates the harvesting of whales.

Water Resources and Pollution

Water Resources

Water is life. The unique characteristics of water support that bold statement. Water molecules are polar. This **polar** nature allows them to cling not only to one another through **cohesion** but also to other molecules through **adhesion**. This also makes water the "universal solvent". Water can also absorb a large amount of heat energy without a significant rise in temperature. This allows it to moderate much of the world's climate. Water covers over seventy percent of Earth's surface. Ninety-seven percent of that is marine or salt water. Glaciers and polar ice account for another two percent. Out of the remaining one percent, less than half is accessible freshwater.

Water is constantly moving through the biosphere through **evaporation, condensation, transpiration** and **precipitation**. These processes are collectively known as the **hydrologic** or **water cycle**. When disturbance is kept to a minimum, the water cycle provides man with a constant renewable resource. However, it has been stretched to its limits with the explosion of the human population, urban and agricultural development in arid environments, and pollution.

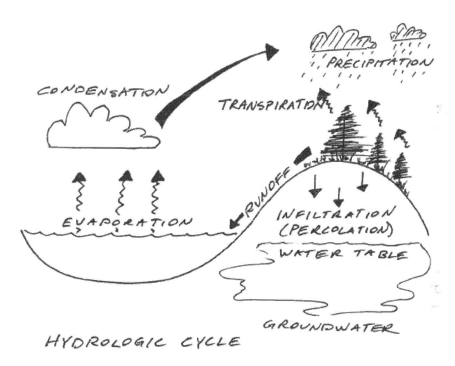

HYDROLOGIC CYCLE

In the United States, water is used for irrigation (41%), energy production (38%), industry (11%) and public use (10%). Streams and lakes account for the water found on the surface while the permeable rock layers of aquifers contain the groundwater.

The illustration below points out the key terms associated with aquifers. The largest aquifer in the United States is the Ogallala Aquifer. It lies on the eastern side of the Rocky Mountains and supplies water to eight states. Agriculture and municipalities rely heavily on this aquifer, which has resulted in lower water tables throughout much of its range. It is estimated that in some arid areas of this aquifer the annual recharge is less than fifteen percent of the use. This not only has obvious implications in urban development and agriculture but it also leads to the subsidence of land.

Methods of irrigation can drastically reduce the amount of water being used by agriculture. **Drip systems** are 90-95% efficient while gravity flow systems are only 60-80% efficient.

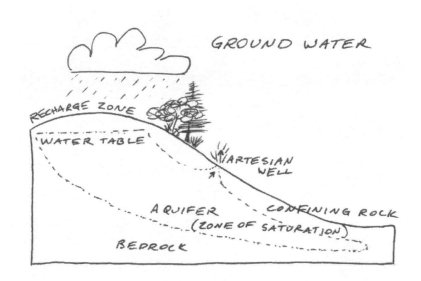

Surface Water

Surface water can be subdivided into moving water (**lotic ecosystems**) and standing water (**lentic ecosystems**).

Lotic Ecosystems

As a river flows from its source to its end, it undergoes a series of changes. In the beginning, the river is typically swift, cold, and oxygen-rich. The river is narrow and very turbulent. Fauna in this area have to be well adapted to this fast moving water. Primary productivity is generally low in this area. As the river distances itself from its source, it enters into a transition zone. The river broadens and production increases. **Turbidity,** or the measure of suspended solids, increases in this zone. While the river continues to slow, it becomes very shallow and broad as it enters into the next zone, the flood plain. Dissolved oxygen decreases, but the area is rich in biodiversity. Wetlands, both inland and coastal, are associated with this zone. As the river empties into the ocean, an **estuary** is formed. These highly productive, biodiverse ecosystems are critical for both terrestrial and aquatic life forms. Estuaries are nurseries for many species of marine fish and offer critical habitat for many migratory birds.

Lentic Ecosystems

Lakes make up the lentic ecosystems. The largest freshwater lake in the United States is Lake Superior while the largest one in the world is Lake Baikal (by volume). The following are the different zones found in lakes.

The **littoral zone** is near the shore. It is shallow enough for roots to gain hold and is highly productive.

The **limnetic zone** (**euphotic**) is the surface of the lake away from shore. Even though it is too deep for rooted plants, there is much production due to phytoplankton and other photosynthetic organisms.

The **profundal zone** is below the limnetic zone and is the area of the lake that does not receive light (**aphotic**). The **benthic zone** is the bottom of the lake. Organisms found here are often dependent on detritus from the limnetic zone.

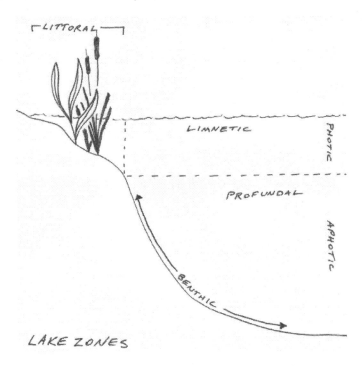

LAKE ZONES

Lakes are often classified by their productivity. **Oligotrophic** lakes are typically young lakes with low productivity. They have clear water, and since they are often associated with colder climates, the dissolved oxygen is typically high. **Eutrophic** lakes are considered mature lakes. Productivity is high and the clarity of water is poor. These lakes are very biodiverse and dissolved oxygen levels are usually lower because of the high demand. **Mesotrophic** lakes fall in between the other two. Although eutrophication is a natural process, man often speeds up the process. When this occurs, it is called **cultural eutrophication**. This is generally a result of a limiting nutrient entering the lake. Typically this is from a nonpoint pollution source like nitrogen fertilizer from agricultural runoff. The increase in the limiting nutrient allows producers to quickly grow and reproduce. **Algal blooms** are often a result. The problem comes when the nutrient is used up and the algae die. Decomposition of the algae reduces dissolved oxygen, which often results in fish die-off.

Lakes are often man-made and these reservoirs provide people with much needed resources. These reservoirs provide water for communities, irrigation, and flood control. They are also a source for recreation. One of the greatest advantages of building these reservoirs is the production of hydroelectric energy. This is a renewable, clean source of electricity. The disadvantages of these projects are loss of biodiversity due to habitat loss, reduction of nutrients deposited downstream, and loss of water through evaporation. The largest reservoir in the United States is Lake Mead created by the Hoover Dam on the Colorado River. Water control projects on the Colorado River have so greatly affected this river, that the once largest river of the western United States, only trickles into the Pacific Ocean. Three Gorges Dam of China and the James Bay Project of Canada allow those two countries produce the greatest production of hydroelectric energy.

Water Pollution

Pollution, of course, affects both flowing and standing water sources. Pollution can be placed in two general categories: **point source** and **nonpoint source**. Point source pollution is easily identified and can be attributed to one source. An example of this would be a pipe from a factory discharging pollution into a local body of water. Nonpoint pollution comes from a broad area. Examples of nonpoint source pollution are agriculture runoff and storm runoff from urban areas. Water can be polluted by nutrients, organic waste, suspended solids, and excess heat. **Cultural eutrophication**, described earlier, is often a result of excess nutrients. **Biological oxygen demand (BOD)** increases with the presence of organic waste since decomposition, which requires oxygen, leaves less oxygen for the inhabitants of the ecosystem. An **oxygen sag curve** (found on the next page) represents the effect that these oxygen demanding wastes have on a stream. Suspended solids can reduce photosynthesis in a body of water disrupting the flow of energy through the food chain. **Thermal pollution**, often due to the cooling process involved in electricity production by nuclear means, often lowers dissolved oxygen and biodiversity.

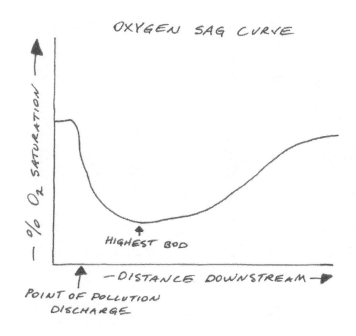

OXYGEN SAG CURVE

% O₂ SATURATION →

HIGHEST BOD

← DISTANCE DOWNSTREAM →

POINT OF POLLUTION DISCHARGE

Pollution can be monitored by a variety of means. Tests are used to determine the chemical composition of water. These tests include **pH**, **dissolved oxygen**, **nitrates** and **phosphates**. **Secchi disks** are used to measure the effect of solids on the clarity of the water. Biological tests can also be used. The biodiversity of aquatic **macroinvertebrates** is a good indicator for water quality.

The Fine Print

Clean Water Act: The CWA regulates the discharge of pollutants into U.S. waters, and sets water quality levels for contaminants in surface waters.

Safe Drinking Water Act: Protects the quality of bodies of water, above or below ground, that are used for drinking water.

Aquatic Biomes

Aquatic biomes are defined based on the presence or absence of salt. **Freshwater biomes** have a salt concentration of less than 1 %; whereas **marine biomes** have a salt concentration greater than 3 %.

Marine Biomes

Oceans cover more than 70% of Earth's surface. Ocean currents, like the Gulf Stream of the Atlantic Ocean, moderate countries' climates by redistributing heat throughout the world.

Vertically, the ocean can be divided into three zones. The **intertidal** or **littoral zone** is the shore line that is covered by water at high tide and exposed at low tide. Barnacles, mussels and hermit crabs thrive here due to their tolerance of harsh, wide-ranging conditions. The **neretic zone** starts outside of the intertidal zone and stretches to the edge of the continental shelf. This area contains ninety percent of ocean life, but only accounts for ten percent of the ocean's area. The open waters past the continental shelf make up the **oceanic zone.** This area is much less productive per unit area than the other zones. However, it accounts for most of the photosynthetic production on Earth due to its vast area. **Nektonic** or strong-swimming organisms are abundant in the oceanic zone.

The ocean can also be divided horizontally into the **euphotic**, **aphotic**, **benthic**, **abyssal** and **hadal** regions.

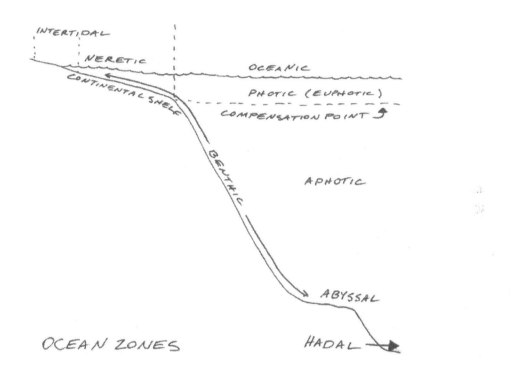

Estuaries, mangrove swamps, coastal marshes, coastal seas, coral reefs, and the open ocean make up the unique life zones found in the ocean.

Estuaries occur where rivers empty into the ocean. This mixture of freshwater and saltwater creates a brackish environment that varies in temperature and salinity. Due to the constant supply of nutrients flowing into this area from the river, this ecosystem is highly productive. These conditions create nurseries for marine fish and provide critical habitat for migratory birds.

Chesapeake Bay is the largest estuary in the United States. In the early 1980s, seafood harvests began steadily declining. This is attributed to the pollution from the more than fifteen million people who live in the watershed of this estuary. The Chesapeake Bay Program was established to restore the bay, and even though it has made much progress, shellfish harvests are a fraction of those just fifty years ago.

Pollution from industrial waste, over-fishing, introduction of invasive species, and urban development have all greatly affected estuaries. Agriculture runoff has led to **cultural eutrophication**. In the Gulf of Mexico this has also led to **hypoxia** (reduction of dissolved oxygen) which has resulted in **"dead zones"**.

Mangrove swamps are coastal wetlands found in the tropical areas of the world. These salt-tolerant trees grow in thick forests along the shore. Their characteristic long roots, used for stability and oxygen attainment, help reduce erosion due to wave action and surges generated by tropical storms. These mangrove swamps also provide a "nursery" for many invertebrate and fish species. This makes these areas ideal sources of fish and shellfish for people.

Mangrove swamps are under immense pressure due to urban and recreational development. The conversion of mangrove swamps for shrimp farms has also led to loss of large tracts of these coastal wetlands. Approximately half of the mangrove swamps in the world have been destroyed.

Coral reefs are found throughout the tropics and subtropics in shallow, clear water. High productivity in these areas supports many species. The very diverse nature of these ecosystems has led to them being called the "rainforests of the seas". The calcium carbonate secreted by corals to form their exoskeletons is the foundation of these ecosystems. Coral reefs are nurseries for many species of commercial fish. They also protect the coastline by absorbing the energy generated by waves and storms.

However, coral reefs are under attack! Pollution, destruction for shipping channels, and damage caused by divers are leading to the decline of healthy coral reefs. Another problem is a phenomenon called **coral bleaching.** This occurs when a section of reef loses photosynthetic algae and is left without a primary producer. Food chains begin to falter and it is left "bleached" white. This is happening in many parts of the world due to an increase in water temperature attributed to global climate change.

Terrestrial Biomes

Both latitude and altitude determine the distribution of heat and precipitation on land. Temperature and precipitation, in turn, determine the distribution of flora. And flora dictates the distribution of fauna. The relationships between these abiotic and biotic factors form the terrestrial biomes of the world.

Tropical Rainforests

Beginning at the equator, where the amount of sunlight and temperature are constant throughout the year, rainforests occur. Tropical rainforests receive over eighty inches of rain a year. The constant presence of sunlight and moisture makes the rainforest very productive. This productivity allows for over fifty percent of the world's terrestrial species to be found in only three percent of the

Earth's land. Due to the high rate of decomposition and constant leaching, the soil of this biome is surprisingly very poor. Nutrients are simply recycled too quickly to stay in the soil. Tropical rainforest trees are typically broad-leaved, but are evergreen. There are dozens of different tree species per acre. Mature tropical rainforests typically have three layers of vegetation. The tallest layer consists of emergent trees, whose crowns stand alone above the rest of the forest. The second layer forms the canopy. This layer is a dense, blanketing layer that is about one hundred feet above the floor. The canopy is home to the majority of life found in the rainforest. The third layer is the understory and is dominated by shade-tolerant plants. Due to the intense competition for sunlight, epiphytes, like bromeliads, are abundant in the rainforest. These plants gain access to sunlight by living in the forks of branches and wedged in the cracks of bark of taller trees. These relationships are commensalistic since they are not robbing nourishment from their hosts.

Due to population growth and industrialization of developing countries, deforestation of tropical rainforests is occurring at an alarming rate. **Subsistence agriculture**, where families produce enough food for themselves, accounts for the majority of the destruction. Farmers cut the trees down, burn them, and then plant crops in the ashes. This practice is referred to as **slash-and-burn agriculture**. Commercial logging and cattle-ranching for exports are also threatening the remaining rainforest.

Tropical Grasslands

Tropical grasslands also occur near the equator. However, they receive much less rainfall. They receive thirty to sixty inches of rain annually. Although this amount of rainfall is not enough to support the growth of trees found in the rainforest, in some areas of the world it does allow for wide-spread, arid tolerant species to exist.   Tropical grasslands with these types of trees are called **savannas**. Savannas support the awesome herds of zebra, antelope, and wildebeest of Africa.

Due to the conversion of these grasslands into domestic animal rangeland, native fauna is decreasing.   Due to overgrazing in these areas, desertification has become a major problem.

Deserts

Deserts cover one-third of Earth's land.   Deserts typically receive less than ten inches of precipitation a year and have extreme temperature changes daily.   Soil is practically nonexistent in deserts; instead it is dominated by sand and gravel. The vegetation found in deserts has to be adapted to a life of extreme temperatures and prolonged periods without water.   Plants in the desert often have thick waxy cuticles, reduction of leaves, and photosynthetic stems.   Many plants often have a reduction of stomata to minimize transpiration.   The annuals of the desert may remain dormant for years and then cover the desert floor in a carpet of colors when rain finally arrives.   Animal life of the desert is dominated by reptiles and rodents.   Small, nocturnal animals are better fit to survive the extreme heat of the desert day.   During times of prolonged heat, these animals often **aestivate** or remain inactive until conditions improve.

In North America, deserts have been damaged by the use of recreational vehicles. The lack of rain makes the recovery of a desert ecosystem almost nonexistent. Urbanization of desert areas has also depleted aquifers to critical points in both Arizona and New Mexico.

Temperate Forests

Much like tropical grasslands, temperate forests occur in areas that receive thirty to sixty inches of precipitation annually. However soil conditions and seasonal variation allow deciduous trees to be abundant in this biome. Unlike evergreen trees, deciduous trees shed their leaves. The plant life here is not nearly as diverse as the rainforest. There may be more species of trees in one acre in the rainforest, than in an entire temperate forest. These broad-leaved trees provide excellent forage for large herbivores, like deer, which in turn attract large predators, like mountain lions. The rate of percolation and evaporation minimize the leaching of nutrients. This along with a steady rate of decomposition makes for excellent soil in this biome.

The deciduous forest of the eastern United States is one of the largest temperate forests in the world. This forest like many other temperate forests has been greatly disturbed for agriculture, logging, and recreation. **Secondary succession** occurs when abandoned farm land is allowed to return to deciduous forest. When this occurs, many native species have successfully reestablished populations.

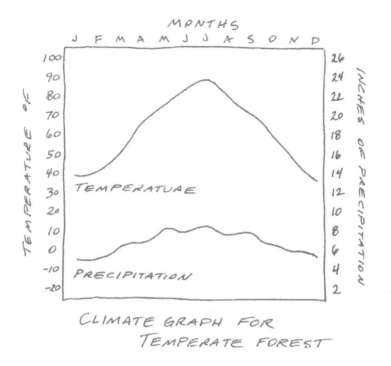

MONTHS

CLIMATE GRAPH FOR
TEMPERATE FOREST

Temperate Grasslands

Temperate grasslands occur in temperate areas of the world where precipitation falls between ten and twenty inches. In South Africa, temperate grasslands are known as the veld. In North America, they are referred to as the Great Plains. In Asia, they are called the steppes. And in South America, they are the pampas. These biomes have provided a vast, reliable food source for the largest herds of hoofed mammals on Earth. The soil of temperate grasslands is rich with organic matter. Grasslands are very resilient biomes and were maintained in the past with frequent fires and grazing.

In the United States, less than one percent of native grassland exists. Once people moved west, settlers discovered that this biome was excellent for agriculture. They replaced the native grasses with domesticated grasses like corn and wheat. Many of the native herbivores, like bison, were **extirpated** and replaced with cattle.

Taiga

South of the Arctic Circle lays a belt of evergreen trees called the taiga. These northern coniferous forests are also called **boreal** forests. Due to its northern location, winter offers only seven hours of sunlight, while summer offers up to nineteen. Taiga is derived from a Russian word that means "swamp forest". This describes the abundance of water, but typically it is locked up in ice and in deep glacier-gouged lakes. These forests are dominated by a few species of evergreen cone-bearing trees. The trees include spruce, fir, and pine. The soil called **podzol** is very poor. Decomposition of the conifer needles is slow and leaves the nutrient-poor soil very acidic. This and a freeze line that extends five feet below the surface further reduce the floral diversity. Fauna include moose, wolves, and martens. Summer months are active with migratory birds and abundant insects.

Tundra

The tundra is located north of the tree line and south of the arctic polar ice cap. This biome has sunless winters and very little falling precipitation. Due to the extreme cold conditions, a frozen layer of subsoil exists year round. This nutrient-poor soil is called **permafrost**. The frozen soil, extreme cold temperatures, and howling winds limit plant life to low growing plants. These include mosses and grasses and occasionally shrubs like dwarf willows. Prolific insect life during the summer beckons migratory birds. Wildlife found year-round includes lemmings, arctic fox, and snowy owls. Typically the animals have abbreviated appendages and small surface area to volume ratio to minimize heat loss. Animals also cope with the extreme winter by either hibernating or migrating.

As we travel from the equator to the poles, we find a familiar trend. Generally we find the same decrease in temperature and precipitation that we find with an increase in altitude.

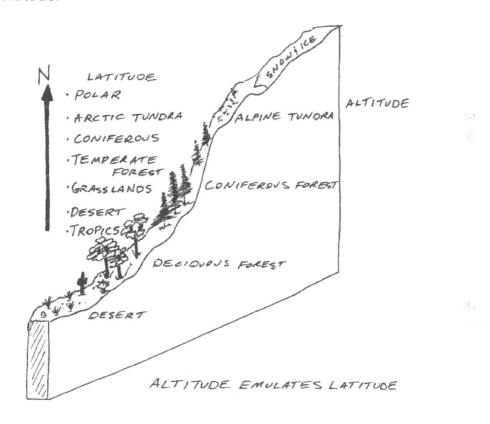

ALTITUDE EMULATES LATITUDE

Geology

Earth is approximately 4.5 billion years old. Scientists believe that life originated on Earth 3.8 billion years ago. These vast amounts of time are best measured with the geologic time scale. The geologic time scale relates geologic events to important evolutionary events.

A cross-section of Earth would reveal three distinct zones: the crust, the mantle, and the core. The crust is the outermost layer and is composed of 47% oxygen, 28 % silicon, 8% aluminum, and 5% iron. The crust plays a critical part in many biogeochemical cycles and is the source for nonrenewable resources. Below the crust lies the mantle. It consists of iron, silicon, oxygen, and magnesium. The core is the innermost zone. It is believed to have a solid iron and nickel interior surrounded by a liquid iron and sulfur outer core.

The outermost layer of the mantle and the crust make up the **lithosphere**. This layer is rigid and broken into large pieces called tectonic plates. Below these plates is the **asthenosphere**. This thin plastic-like layer of molten rock allows the movement of the tectonic plates and the continents that rest upon them. This phenomenon is known as the theory of continental drift. Volcanoes, earthquakes, mountains, and oceanic trenches are the results of the interaction of these plates.

Plate Boundaries

Convergent boundaries are the result of two plates pushing together. The Himalayas are a result of the Indian-Australian Plate and the Eurasian Plate colliding into one another. When plates move away from one another it is called a **divergent boundary**. One of the best examples of this type of boundary is the Mid-Atlantic Ridge. A **transform fault** occurs when two plates slide past each other in opposite but parallel directions. In the United States, the San Andreas Fault of California is an example of a transform fault.

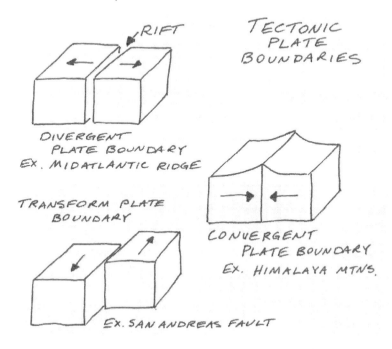

TECTONIC PLATE BOUNDARIES

RIFT

DIVERGENT PLATE BOUNDARY
EX. MIDATLANTIC RIDGE

TRANSFORM PLATE BOUNDARY

CONVERGENT PLATE BOUNDARY
EX. HIMALAYA MTNS.

EX. SAN ANDREAS FAULT

Earthquakes

Earthquakes are violent vibrations that are caused by tectonic activity. The point of the earthquake's origination is called the **focus**. Directly above the focus on the surface is the **epicenter**. Earthquakes are measured by seismographs. Data collected by seismographs are analyzed and reported as a magnitude on the **Richter magnitude scale**. Much like the pH scale, the Richter magnitude scale is logarithmic. **Logarithmic** scales measure values that span a wide range. Moving from one value to the next, on a logarithmic scale, results in a ten-fold increase or decrease.

Volcanoes

Volcanoes are also a result of tectonic activity. They are associated with subduction zones at convergent plate boundaries and divergent plate boundaries. These areas allow molten rock or magma to flow to the surface where it is then referred to as lava. The **Pacific Ring of Fire** is an area of earthquakes and volcanoes that occur around Asia, North America, and South America due to the subduction zones of the Pacific Plate. Hot spots are also responsible for volcanoes. **Hot spots** are the result of plate movement but are not confined to the boundaries of the plates. Volcanoes caused by hot spots form due to an opening or weak spot in the crust that allows a plume of magma to push to the surface. Once the lava breaches the water's surface and islands form, **pioneer species** like lichens and mosses invade the rocky substrate. These organisms begin the process of **primary succession** which forms soil and allows other organisms to colonize. The Hawaiian Islands were formed in this fashion.

Rock Cycle

Tectonic activity, wind, water, temperature, and gravity act together to form different types of rocks. The rock cycle describes this process. Magma solidifies and forms igneous rocks. Igneous rocks are then exposed to the forces of erosion and weathering and break down into smaller particles. These sediments undergo compaction and cementation and become sedimentary rocks. Temperature, pressure, and chemical reactions can change sedimentary and igneous rocks into metamorphic rocks.

Wind

The focus of this portion of the book will be the atmosphere. We will cover the basics of weather, climate change, and air quality.

The Atmosphere

Our atmosphere is a layer of gases that surrounds the planet. These gases are held close to the planet by gravity. For the purpose of the AP exam, you should be concerned with the following layers of the atmosphere. (See Diagram on the following page)

Troposphere: The troposphere is the layer of the atmosphere that is closest to the surface of the earth. The troposphere contains the air that we breathe and live in. The troposphere is made up of approximately 78% nitrogen, 21% oxygen, and small amounts of CO_2, methane, water vapor, and other trace gases.

Stratosphere: The stratosphere is the next layer of the Earth's atmosphere. The stratosphere is important for our purposes as it is the home of the ozone layer. Ozone (O_3) is a gas that blocks a great deal of UV light as it enters the atmosphere.

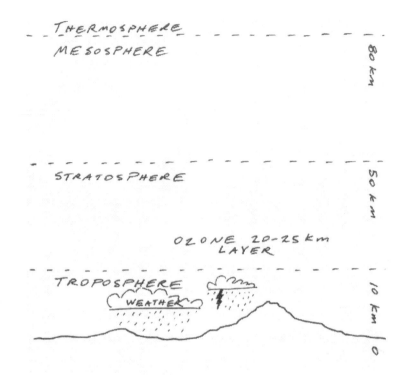

Common Air Pollutants in the United States

Point Source Pollution is a type of pollution that can be traced to a single source. Point source pollution is fairly easy to regulate as it is easy to identify the source. A smokestack on a factory would be an example of point source pollution.

Non-point Source Pollution is caused by multiple sources acting together. This type of pollution is more difficult to control as the source is not as easily identified. Automobile exhaust is a good example of non-point source pollution with many cars each contributing a small amount of pollution.

Air pollution can come in the form of either **primary pollutants**, which are put directly into the atmosphere, or **secondary pollutants** which form from the interactions of primary pollutants with air and sunlight.

Acid deposition is simply the depositing of acid forming compounds on land, or water. Deposition can either be wet (acid rain), or dry (particulates). Acid deposition can disrupt terrestrial and aquatic ecosystems by altering the pH of the soil or water. Acid deposition is best dealt with by reducing the formation of acid forming compounds such as SO_2. In some cases, lakes that have been affected by acid deposition have been treated with large amounts of lime, or limestone, to buffer the acid.

Carbon Monoxide is a colorless, odorless gas that forms during the incomplete combustion of fossil fuels.

Sources: Cigarette smoke, incomplete combustion of fossil fuels, motor vehicle exhaust is the #1 source

Health and Environmental Effects: Affects O_2 carrying capacity of red blood cells contributing to heart attacks, fetal development, pulmonary disease, brain cell damage

Lead is a solid toxic metal that is emitted into the air as particulate matter.

Sources: Paint, metal refineries, lead manufacturing, batteries, leaded gasoline

Health and Environmental Effects: Accumulates in the body causing brain damage, mental retardation, and digestive problems. Lead can negatively impact wildlife. Efforts have been made to remove lead from gasoline and paint but some countries persist in selling leaded gasoline.

Nitrogen Dioxide is a reddish-brown gas that gives smog its brown color. In the atmosphere it can be converted to nitric acid. nitric acid can then fall as acid rain.

Sources: Fossil-fuel burning in cars and power/ industrial plants

Health and Environmental Effects: Lung irritation, NO_2 can increase susceptibility to respiratory problems, and irritate asthma. NO_2 can reduce visibility and contribute to acid deposition. As a component of acid rain it can degrade soil quality, harm aquatic life, and damage buildings.

Sulfur Dioxide (SO_2) is a colorless gas that is generated by the combustion of sulfur containing fuels like coal and oil.

Sources: The incomplete combustion of fossil fuels. Many people are exposed to unsafe levels through improper ventilation of heating and cooking exhaust.

Health and Environmental Effects: SO_2 can irritate asthma and contribute to acid deposition.

Suspended Particulate Matter: SPMs are particles, or droplets, that are light enough to stay suspended in the air for extended periods of time.

Sources: SPMs can be created by burning, plowing, construction, or anything that puts particulates into the air.

Health and Environmental Effects: SPMs can cause respiratory tract irritation. Some SPMs such as lead, dioxin, or PCBs can cause much more serious health problems.   SPMs can reduce visibility and cloud water.   SPMs in water can interfere with the function of fish gills and impact photosynthesis by clouding the water.

Ozone (O₃): Ozone is a highly reactive gas. Ozone days, ozone action days, ozone alert days, or whatever your weatherman calls them, are days when weather conditions are just right for an accumulation of ground-level ozone. Since ozone is formed as a component of photochemical smog, sunny days with little wind are prime days for ozone accumulation.

Sources: Ozone forms in the troposphere as a secondary pollutant (photochemical smog).

Health and Environmental Effects: Ozone is a lung irritant.

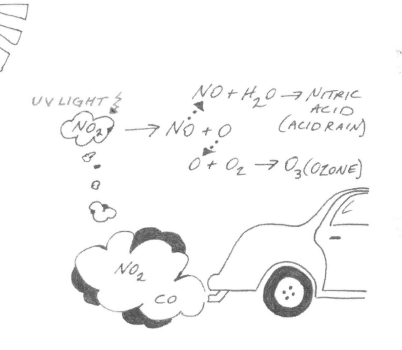

Temperature Inversions can cause air pollution to concentrate and cause increased health impacts. Inversions occur when a warm air mass moves into an area and traps the existing cold air mass below. The stagnant air at ground level concentrates any air pollutants that have been generated in the area.

Prevention of Air Pollution

Air pollution is a particularly interesting problem since air has no boundaries. A pollution source in one place or country can be a pollution problem in another. The problem is further complicated as there are both point and non-point sources of pollution. Air pollution is one of the environmental success stories in the United States. While we still have a long way to go, there have been major improvements in air quality over the last 50 years. Cleaner technology developed for industry and transportation along with air quality legislation and consumer demand has improved air quality. Point sources of pollution have been dealt with through legislation such as the Clean Air Act, and non-point sources, such as automobile exhaust, are generally managed through local requirements for automobile inspections.

The Fine Print

Clean Air Acts 1970, 1977, and 1990

The Clean Air Acts give the Environmental Protection Agency (EPA) the power to regulate emissions of air pollutants.

Cap and trade or emissions trading programs allow companies a certain amount of pollution. If they do not use it, they can trade or sell their allotment to other companies.

Indoor Air Pollution

Indoor Air Pollution is a significant health issue. Air quality inside homes and workplaces is often worse than the air outside. As we have developed our building techniques to make homes better insulated, we have inadvertently made them better at trapping indoor air pollution where we spend most of our time! Indoor air pollution can include gases from lawn chemicals, paints, household chemicals, tobacco smoke, and carbon monoxide. Allergens such as dust, pollen, and smoke particles also contribute to decreased indoor air quality.

Radon is a radioactive gas that can enter homes from the surrounding bedrock. Radon has been implicated in lung cancer cases of people who live in homes with high concentrations of radon.

Weather

Weather is a short term and local phenomenon. It is what is happening in your town at the particular moment. The characteristics of weather include temperature, precipitation, wind direction, barometric pressure, and humidity. The key is to remember that weather is a highly variable local occurrence. An abnormally warm summer or cold winter does not imply a climate shift.

Climate on the other hand, is a pattern of weather that has occurred over a long period of time. The uneven heating of the surface of the earth, ocean currents, and patterns of air circulation will determine the normal temperature and precipitation for a particular area. Precipitation and temperature are the two main determinants of climate. **Microclimates** are small areas that differ from the surrounding climate and can occur in the form of rain shadows and heat islands.

A **heat island** occurs when a city containing large amounts of concrete and pavement stores large amounts of heat. The result is an average temperature that is slightly higher than the surrounding area.

The **rain shadow effect** is described in the diagram below.

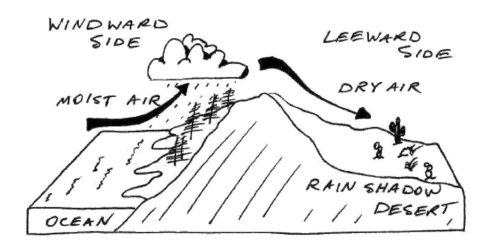

A **warm front** occurs when an incoming warm air mass meets a cooler one that is in place. The warm front rises above the cool air below it and condenses to form precipitation. This type of front can create long periods of clouds and precipitation.

A **cold front** occurs when an incoming cold air mass meets a warm air mass that it is replacing. The cold front stays near the ground, effectively forcing the warm air upwards. As the warm air is forced upward, it creates tall clouds that can generate heavy rainfall, high winds, and thunderstorms.

Ocean currents play a large role in climate patterns as they redistribute heat to different regions.

Extreme Weather

 In some circumstances weather, such as tornadoes and hurricanes, can be quite destructive. **Tornadoes** can form when a cool downdraft and a warm updraft meet to create a spiraling funnel cloud.

Hurricanes (typhoons in the pacific) form in a similar fashion to tornadoes, and are sustained by the energy and moisture from warm ocean water. The shallow Gulf of Mexico is the perfect environment for hurricanes to gain size and strength.

Aside from normal weather variances there are weather cycles that occur infrequently, and have far reaching effects on many regions of the earth. **El Niño - Southern Oscillation** (see diagram on the following page) is an interesting weather pattern that occurs every few years in the equatorial and South Pacific Ocean. The changes in the weather shown in the diagram below can lead to droughts, heavy rainfall, mudslides, and increased incidence of disease in many countries around the globe. The suppression of nutrient upwellings results in decreased biological productivity as nutrient rich water does not make it to the surface waters.

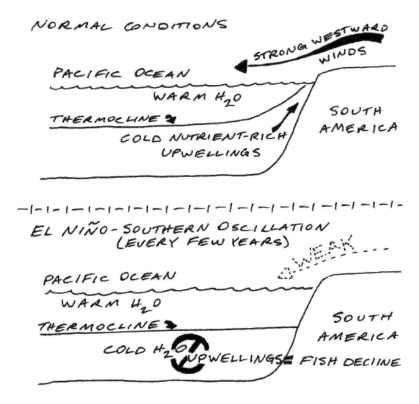

The **Greenhouse effect** occurs when sunlight penetrates the Earth's atmosphere. Some of the energy is released as longer wave radiation from the surface of the earth. This radiation is "trapped" by atmospheric gases such as water vapor, CO_2, methane, and nitrous oxides. This "trapped" energy has a warming effect on the troposphere. The greenhouse effect is necessary for maintaining life as we know it. THIS IS GOOD.

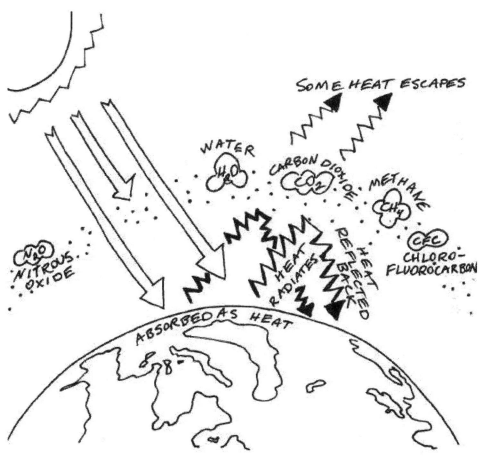

Global Change

Extinction of species is a natural occurrence. It is said that 99% of all organisms that have ever lived are now extinct. The rate at which this natural process of extinction occurs is called the **background extinction rate**. Current extinction rates have been calculated to be 100 times greater than the background extinction rate. This rapid increase in the extinction rate is being referred to as The Sixth Mass Extinction. The sixth differs from the previous five in that they were all natural occurrences. The current extinction rates are driven by increased development, poaching, commercial fishing, and other human activities. As we struggle to maintain our economies and care for our increasing population, the inadvertent side effect is often destruction of habitat leading to a loss of biodiversity.

The Fine Print

The **Endangered Species Act** protects endangered plants, animals, and their habitats. The ESA is a U.S. Law that was established in 1973.

The Convention on International Trade in Endangered Species (CITES) is an international treaty that regulates the trade of species that are threatened or endangered.

Global Warming is simply an increase in the average temperature near the surface of the earth. The more controversial definition of global warming includes the assumption that the recorded warming is driven by anthropogenic causes. The possible connection between human activity and global warming lies in our production of greenhouse gases. It is theorized that CO_2, methane, and other greenhouse gases that are generated from human activities, such as burning fossil fuels and maintaining livestock, are increasing the greenhouse effect to the point that the Earth's average temperature is increasing. This increase in temperature is resulting in the rapid melting of the planet's glaciers and polar ice. This rapid melting could lead to an increase in sea level and a disruption in the ocean currents that drive weather and climate patterns. Ocean currents are often referred to as the ocean's conveyor belt since they carry heat from the equator and redistribute it throughout the earth.

An interesting side story in the global warming saga is that of the Earth's albedo. **Albedo** refers to the reflectivity of the surface of the Earth. Simply put, surfaces like ice reflect more of the energy from the sun while surfaces such as bare rock or soil absorb more of the energy. As glaciers and icecaps recede, more rock and soil will be exposed. This will result in more of the sun's energy being absorbed by the surface of the Earth, further increasing the intensity of the warming effect. This is an example of a **positive feedback mechanism**.

Evidence of Climate Change

Climate change, including global warming, is a complex issue. The work of many scientists from many countries is considered. Complex computer modeling is used as ideas about climate change are formed. One important thing to consider is that climate change means climate change. While some portions of the Earth will become drier others will experience record precipitation. While temperature increases in some areas, it will decrease in others. Climate change is a complex issue that cannot be simplified into a brief discussion. Without jumping into the debate on climate change, the following list represents some of the conclusions of research that support the theory that the Earth is currently undergoing rapid change.

1. Since the beginning of the industrial revolution, atmospheric concentrations of CO_2, methane, and nitrous oxide have increased significantly.

2. Due to the sheer volume of CO_2 that is generated, CO_2 is the greatest anthropogenic contribution of greenhouse gases.

3. Most research indicates that human contributions with a warming effect are greater than human contributions that have a cooling effect on the climate.

4. Milankovitch cycles are changes in the Earth's rotation and orbit around the sun. Throughout the history of the Earth, these cycles have impacted climate change.

5. Solar output varies and has some regular cycles of increased and decreased output. Changes in solar output can account for only a small warming effect, not one of enough magnitude to account for observed warming.

6. CO_2 concentrations have increased dramatically since the 1950s.

7. Since the early 1900s, the Earth's average temperature has risen by almost 1 degree Celsius.

8. Glaciers and sea ice have been melting at increasing rates. Since 1979, more than 40% of the polar ice cap in the northern hemisphere has receded.

9. Precipitation trends are changing with some areas becoming wetter and some areas becoming drier.

10. Sea levels have risen slightly. This is partially due to the **thermal expansion** of the ocean.

11. **Coral bleaching** occurs when temperatures rise and cause the symbiotic algae to leave the coral. The coral takes on a white appearance in the absence of the algae and often dies shortly after.

12. The ocean absorbs CO_2. The CO_2 forms an acid in the water causing the pH of the ocean to decrease. As atmospheric concentrations of CO_2 continue to rise this process has accelerated and is causing the ocean pH to decrease further. This process is called **ocean acidification**. This decrease in pH could impact the ability of coral and other organisms that form calcium carbonate shells to exist.

13. The number of frost free days per year is increasing throughout the United States. This is impacting the natural cycles of plants. This could lead to a mistiming between the flowering of plants and the seasonal arrival of the plants' pollinators.

14. In the last fifteen years, Plant Hardiness Zone maps have been redrawn to accommodate changes in temperature and precipitation in the United States.

15. In the western United States, fire season has increased significantly over the last 20 years, and the number of large fires has increased. This is most likely due to changes in precipitation. Forest fires are also occurring at higher altitudes than previously recorded. This is most likely due to decreased snow pack in the mountains.

16. The ranges of some pest species have expanded as average temperatures have increased. This could result in dramatic losses for agriculture, and the increased incidence of disease as vectors, such as mosquitoes, are able to thrive in new areas.

While it appears that there are significant changes occurring in our atmosphere we are left with two simple, but very important, questions. What does climate change mean for us, and what, if anything, can we do about it? First we will address some of the potential effects of a changing, or warming, climate.

Potential Disadvantages of a Changing Climate

Ice caps, glaciers, and permafrost will continue to melt. The large amount of CO_2 that is currently sequestered in the permafrost could be released into the atmosphere.

As regional climates shift, living organisms will struggle to adapt to their new environments. This will almost certainly lead to a further decrease in biodiversity.

It is possible that the size and strength of storms in some areas will increase in a warming climate.

A warming climate will lead to new health concerns as diseases, and their vectors, are able to spread into parts of the globe where cooler climates previously prevented their infiltration.

As climate shifts occur, it will become more difficult to produce crops in areas that currently supply large portions of our food. Conversely, there will be some areas that are productive where they were not before.

What Can We Do About It?

Here are some potential actions that could be taken to reduce the impact of climate change.

1. Continue to improve the efficiency of our homes and transportation. More efficient living means less fossil fuel consumption and the emissions that go along with it.

2. Continue the shift to established renewable energy resources such as wind and solar power.

3. Diversify energy production by using local resources to produce electricity.

4. Increase the efficiency of existing coal-fired power plants.

5. Reduce deforestation. Cutting down trees not only removes the trees carbon sequestering ability, but prematurely releases the carbon that was stored in the tree.

6. Use more natural gas to replace coal in energy production.

7. Use conservation tillage on a greater percentage of croplands.

8. There are several options for sequestering CO_2 that is already in the atmosphere. These methods include planting more trees, injecting CO_2 into the deep parts of the ocean, and chemically removing CO_2 from automobile and factory exhaust.

The Fine Print

The **Kyoto Protocol** is an international treaty that is designed to reduce the emission of greenhouse gases with the goal of reducing the impact of climate change. The U.S. initially signed the treaty, but then failed to ratify the treaty in the U.S. Senate.

The **Intergovernmental Panel on Climate Change (IPCC)** is an international panel whose task is to evaluate the impact of climate change. The panel is responsible for the IPCC report which is an ongoing evaluation of climate change.

Ozone Depletion

The ozone layer in the stratosphere helps block UV light from reaching the surface of the Earth. This is particularly important because UV light is detrimental to living organisms. Increased UV radiation, resulting from a thinning ozone layer, has been implicated in increases in human skin cancer, and in the decline of amphibians worldwide. There is a seasonal thinning of the ozone layer that occurs over the poles. The thinning over Antarctica is the most significant. This seasonal thinning is not what we are primarily concerned with when we discuss Ozone depletion.

A group of chemicals called chloroflourocarbons (CFCs), along with other ozone depleting chemicals (ODC's), has slowly been breaking down the stratospheric ozone layer since CFCs were first put into use prior to World War II. CFCs were widely used as propellants and refrigerants in everything from hairspray cans to automobile air conditioners. CFCs are persistent in the environment. Other ODCs include methyl bromide, hydrogen chloride, and halons.

UV light reacts with CFCs to break a chlorine molecule off of the CFC. The chlorine then "steals" an oxygen from ozone (O_3) The chlorine binds with the oxygen leaving O_2 which does not block UV light! Free oxygen atoms then "steal" the oxygen from the remaining CLO leaving chlorine alone to go steal another oxygen from an ozone molecule.

Negative Impacts of Ozone Depletion

With a thinner ozone layer, more UV light will pass through the atmosphere resulting in increased incidence of skin cancer, cataracts, and sunburn.

There will be increased incidence of photochemical smog as more UV light is available to react with fossil fuel emissions.

Some populations of aquatic organisms particularly susceptible to UV radiation will experience a decline in numbers.

The Fine Print

In 1987 the **Montreal Protocol** was passed. This is an international treaty to drastically reduce CFCs prior to the year 2000.

In 1992 The **Copehhagen Protocol**, another international treaty, was adopted. The goal of the Copenhagen Protocol was to phase out other Ozone Depleting Chemicals as well as CFCs.

The two treaties have played a major role in the reduction of ODC emissions.

Fire

Fire is simply the oxidation of combustible materials that generates light and heat. What an appropriate place to review the sources of energy that make life as we know it possible! Here we will cover the basics of production and use of energy. We will take an in-depth look at nonrenewable, renewable, and alternative energy sources. We will consider the economic and environmental impacts of our major sources of energy.

Non-Renewable Energy

Nonrenewable energy resources, such as coal, oil, and natural gas account for the vast majority of the world's commercial and transportation energy. These fuel sources provide relatively inexpensive, high-quality energy. Yet, they are also responsible for several major environmental problems. By definition, a nonrenewable resource is one that will eventually run out. Here we will take a look at the major sources of nonrenewable energy, their advantages, and their disadvantages.

Coal

Coal was formed when organic matter (mostly from plants) was exposed to increasing heat and pressure underground. Over long periods of time the organic matter is transformed into the compact carbon based matter that is coal. There are several types of coal that are formed depending on the amount of heat and pressure that the material is exposed to.

Peat: Peat is an early stage of coal that is formed under minimal pressure and has high moisture content. Peat has a very low heat and carbon content. Peat is harvested and burned for heat and cooking. It is found predominantly at higher latitudes in the northern hemisphere.

Lignite: Lignite is coal that has been formed under more pressure, heat, and time than peat. Lignite still has a relatively low heat and carbon content. It also has high moisture content. Lignite is burned to produce electricity, but is responsible for a great deal of air pollution due to its high sulfur content.

Bituminous Coal: This type of coal is used regularly as a fuel because of its high heat content. It also has high sulfur content.

Anthracite (hard coal): Anthracite is a very valuable energy resource because it has a high heat content with very low moisture and sulfur content. Anthracite takes much longer to form than the other types, and is more expensive.

Minerals such as coal are removed by either surface or subsurface mining techniques. **Subsurface mining** is used when the mineral deposit is too deep to extract using surface mining techniques. Surface mining techniques include:

Open-pit mining: Large holes are dug directly into the surface of the earth to remove ores.

Strip mining: Heavy machinery is used to dig large trenches. The minerals are removed and the overburden is used to fill in the previous trench.

Contour strip mining: The same concept as strip mining is applied in mountainous areas.

Mountain-top removal: Heavy machinery or explosives are used to remove the top of a mountain in order to expose the minerals.

Impacts of Mining

Surface mining disrupts habitat which can lead to loss of biodiversity. Disturbed soil is also more susceptible to erosion. Subsurface mining can lead to **subsidence** as the ground above the mine caves in over time. Fires can start in underground coal deposits, and burn out of control for long periods of time. Acid mine drainage can occur when rainwater leaches through mine waste, and causes acid to runoff into nearby bodies of water. Old mining sites can be restored or reclaimed by replanting native species of plants on the disrupted land. All too often mining companies go bankrupt or shut down operations without restoring the disrupted land.

Coal Burning Power Plants

Coal is used throughout the world to directly heat homes, heat water, cook, and generate electricity. Approximately 40% of the world's electricity is generated by burning Coal. Electricity is produced from coal by the following process:

Coal is pulverized and then burned. The heat is used to boil water in a boiler. The steam that is generated is used to spin a turbine which in turn generates electricity. The steam is then condensed and the water is returned to the boiler to be used again. Waste heat generated during the process can be transferred to the atmosphere through a cooling tower or to a body of water. On a side note, it is possible to use the waste heat for commercial or residential hot water through a process called **cogeneration**.

Good reasons to use coal!

Coal is an abundant resource in the US with a relatively high net energy yield. The infrastructure for mining and using coal is well developed, and it is relatively inexpensive to use.

Negative impacts of using coal

A great deal of CO_2 is released when coal is burned. The burning of coal also releases SO_2 which is involved in the formation of acid rain.

Crude Oil

Crude oil is a fossil fuel that is produced by the decomposition of deeply buried dead organic matter. These plants, animals, and protists were exposed to high pressure and temperature over millions of years. The resulting product is crude oil that is composed of hydrocarbons and small amounts of sulfur, oxygen, and nitrogen. Crude oil is extracted by drilling holes into a deposit and pumping the oil to the surface for distribution. Petrochemicals are used not only for fuel, but in the production of plastics, asphalt, pesticides, grease, wax, and industrial solvents. A process called **fractional distillation** is used to separate the crude oil into the gas, jet fuel, heating oil, solvents, etc. that we use today.

Oil is now the largest source of commercial energy in the world. The OPEC countries hold as much as 60% of the worlds proven crude oil reserves. The three largest consumers of oil are the United States, China, and Japan.

There are many advantages to using oil as a fuel source such as:

Oil is relatively inexpensive. It is easily transported with pipelines, trucks, and ocean tankers. There is relatively low land use associated with drilling for oil. Oil has a relatively high net energy yield.

Some disadvantages to using oil as a fuel source include:

Demand for oil may exceed production in the next 50 years. Throughout the life cycle of the production and use of oil it causes environmental degradation. Drilling for oil causes land disruption which can lead to rapid soil erosion. Oil spills can be very disruptive to aquatic habitats. Petroleum products can contaminate groundwater supplies. Burning oil for energy produces CO_2, a greenhouse gas, as well as sulfur dioxide and nitrogen oxides which lower air quality and contribute to acid rain.

The Fine Print

The **Oil Pollution Act of 1990** gives the EPA the ability to respond to oil spills and to regulate oil storage facilities.

Shale Oil and Tar Sands

Shale oil is extracted from shale rock that contains a mixture of hydrocarbons called kerogen. The rocks are crushed and then heated to separate the shale oil from the rock. The majority of the world's shale oil can be found in the United States. Unfortunately it takes a great deal of energy and water to extract the shale oil. Tar Sand is a thick substance made of sand and bitumen. The majority of the world's tar sand can be found in Alberta, Canada. The tar sand must be heated significantly to extract the oil. Both shale oil and tar sands require a great deal of land disruption during the mining process. Shale oil and tar sands have a lower net energy yield than conventional oil and share the same environmental concerns when used as a fuel source.

Natural Gas

Natural gas is a mixture of methane, butane, ethane, and hydrogen sulfide. Natural gas is found above conventional oil deposits. A pipeline must be in place to trap the natural gas. Gas is often burned off as a waste product in oil production. Components of natural gas are compressed and stored as liquefied petroleum gas. The natural gas that is sent through a pipeline to be used for heating and electricity production is methane. Natural gas can be refrigerated and liquefied for transport over long distances.

Some benefits of natural gas include:

Natural gas has a relatively low cost with a high net energy yield. There are plentiful reserves of natural gas in oil producing countries. Natural gas has lower carbon dioxide emissions than conventional oil and contributes less air pollution.

Some disadvantages of natural gas include:

While it is better than conventional oil, natural gas is still a nonrenewable resource that produces CO_2 and air pollution when it is used. Damaged pipelines can release methane, which is a powerful greenhouse gas, into the troposphere.

Nuclear Power

Nuclear power is considered an alternative energy source as is does not directly use fossil fuels. However, nuclear power uses fissionable materials such as Uranium-235 which has to be mined, and is technically a nonrenewable resource.

To produce electricity, controlled nuclear reactions are used to generate a tremendous amount of heat which is used to create steam which spins a turbine. The spinning turbine generates electricity, and the waste water must be cooled and condensed much like a coal power plant. Nuclear power currently accounts for only a small percentage of the world's electricity production, and that percentage is actually predicted to decrease as the rate of decommissioning of old nuclear power plants exceeds the development of new nuclear power plants. Nuclear fission, the splitting of atoms, is responsible for generating the heat that is used in a nuclear power plant. Experiments have been done with hopes of using nuclear fusion for energy production, but attempts have been unsuccessful.

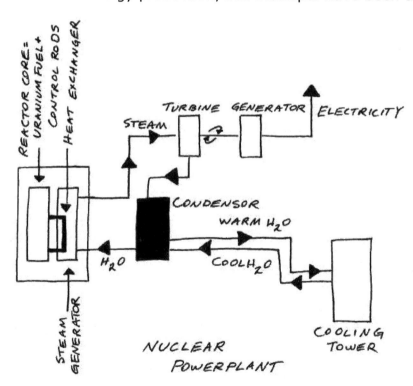

Radioactive material such as Uranium-235 decays into other lighter radioisotopes. As the isotopes decay, they give off radiation and sub-atomic particles. The rate at which the isotope decays is its **half-life**. Half-life is the amount of time that it takes for half of the atoms in a given amount of radioactive material to decay. Uranium-235 has a half-life of 700 million years.

Yucca Mountain in southern Nevada has been chosen as a central site for nuclear waste disposal in the United States. If approved, radioactive waste from the military and power plants would be stored underground at the Yucca Mountain site indefinitely. Chernobyl (Ukraine 1986) was a large nuclear disaster caused by a meltdown in a nuclear power plant. Human error was the main cause cited in the disaster. Three Mile Island (Pennsylvania 1979) was a nuclear incident involving a partial core meltdown.

Benefits of using nuclear Power

Using nuclear power does not directly contribute to greenhouse gases or air pollution.

There is relatively little land disruption.

There is a relatively large supply of nuclear fuel.

Nuclear power plants are heavily scrutinized and relatively safe.

Some disadvantages of nuclear power include:

Nuclear waste is extremely dangerous and very expensive to dispose of.

Nuclear power plants serve as potential terrorist targets.

A tremendous amount of waste heat is generated by the power plants.

While the risk is low, nuclear disasters are very dangerous.

Renewable Energy Resources

Passive Solar

Passive solar technology directly captures sunlight and converts it into heat. Sunlight can be used to either directly heat air or water. Passive solar design can be incorporated into buildings or stand alone devices can be built. In a home, passive solar can be employed by allowing sun to enter large south-facing windows to warm the house during winter time. This effect can be magnified by incorporating "thermal mass" into the building design. This means that the builder would use stone, water, etc. to store heat when the sun is shining and later release heat when the sun goes down.

Advantages of Passive Solar

It is a highly efficient use of solar energy.

There is relatively low land disturbance.

No fossil fuel is used; therefore, no pollution is associated with fossil fuels.

Disadvantages of Passive Solar

Initial construction and design can be costly.

Passive solar works best with clear sunny skies.

Like Ants Under a Magnifying Glass....

In some solar applications the sunlight is magnified using mirrors to direct the energy into a single area. Like a large solar cooker, this method can be used to produce steam for electricity production.

Photovoltaic Cells

PV cells, commonly called solar panels, collect sunlight and convert it to electricity. Sunlight causes the release of electrons in the panel that creates a current. PV Cells can be used to charge batteries allowing the electricity to be stored for later use or they can supply electricity directly to a motor, light, etc. PV cells can be used on a large scale to provide electricity to buildings or in small applications such as solar garden lights or solar calculators. Most of the southwestern United States receives enough solar radiation to make PV cells a viable option for electricity production.

Advantages of Photovoltaic Cells

No fossil fuels are used, and the air pollution associated with fossil fuels is avoided.

No Greenhouse gas emissions are produced.

They are perfect for remote areas where the infrastructure for power may not exist.

Once the panels are installed, maintenance is low and the electricity is free!

There are no moving parts in a solar panel which leads to less repairs and lower maintenance.

Disadvantages of Photovoltaic Cells

Electricity is only produced when the sun is shining.

Energy is consumed in the production of solar panels.

Pollution is generated in the manufacturing and transportation of solar panels.

The initial cost of solar panels is high.

Outdoor solar panels must be cleaned regularly to maintain efficiency.

Wind

Wind power is the fastest growing source of renewable energy. Windmills have been in use for hundreds of years, harnessing the power of the wind to pump water or power machinery. The modern windmill uses the wind to turn a generator that spins and produces electricity. These wind turbines can be used in isolation for small applications, like single households, or can be linked together in wind farms that are capable of producing considerable amounts of electricity for residential and commercial use.

Advantages of Wind Power

There are many places throughout the U.S. and the world that have winds that are suitable for energy production.

Once the wind turbine is in place, the electricity is free and renewable.

No greenhouse gas emissions, or air pollutants, associated with burning fossil fuels.

Disadvantages of Wind Power

Wind turbines require a costly initial investment.

Wind turbines only produce electricity while the wind is blowing.

Windmills can be disruptive to migratory birds, and have been shown to kill both birds and bats.

Some consider wind farms to be visually unappealing.

Wind turbines require routine maintenance.

Biomass

Biomass energy production is perhaps the oldest and most widely used method of producing energy for human use. Biomass includes burning firewood, dung, or other waste materials. It also includes decomposing organic material to produce methane and burning plant material to produce steam for electricity production. In its simplest form, wood and dung have been burned to provide heat. In more modern applications, animal waste can be decomposed in large digesters where the resulting biogas can be used for heating, cooking, or electricity production.

Advantages of Biomass Energy Production

Fuel is abundant and locally derived. Wood can be used in wooded areas, dung can be used near livestock, grasses can be used if available, etc...

Waste from sewage plants or feedlots can be used to produce electricity.

Burning biomass releases recently sequestered carbon as opposed to burning fossil fuels which release carbon that has been sequestered for thousands of years.

Fuel is relatively inexpensive.

Waste from industry, such as the sludge left over from beer making, can be digested to produce biogas.

Disadvantages of Biomass Energy Production

If not managed properly, resources like forests can be overused for fuel wood and charcoal production.

The energy content of biomass is relatively low when compared to fossil fuels

Biodiesel / Ethanol

Ethanol is produced by fermenting sugar or starches. When fermentation is finished, the resulting liquid is distilled to make ethanol. Ethanol makes up a small percentage of domestic fuel use. The good thing about ethanol is that it is produced locally which reduces foreign oil imports. Unfortunately, ethanol production involves large amounts of cropland which are already stressed for food production. As more corn and other agricultural products are purchased for fuel, the price for corn-based food products also increases. Ethanol produces less air pollution than its fossil fuel counterparts, but is also very energy intensive to produce. Ethanol will continue to be used as a fuel additive, and for special applications, but it is highly unlikely that it will ever represent a major portion of U.S. fuel consumption.

Biodiesel is produced by mixing plant oil with an alcohol. The resulting biodiesel runs in any diesel engine. Biodiesel has great promise for small scale local fuel solutions because almost any type of plant oil can be used to produce it. Recycled vegetable oil from restaurants can be used to make biodiesel, as long as you don't mind your car exhaust smelling like French fries!

Moving Water Energy Production

For ages, moving water has been used to turn waterwheels which drove machinery. This same concept can be applied to the production of electricity. Hydroelectric power is generated by flowing water turning a turbine that is connected to an electrical generator. As the generator spins, it produces electricity. In some cases, fast moving rivers can be harnessed to produce electricity. More often than not, a dam is built on a river to create a reservoir on the upstream side of the dam. The trapped water is then allowed to flow through the generators.

Advantages of Hydroelectric Power

When a dam is built, land is flooded upstream to create the reservoir for drinking water and recreation.

Hydroelectric power does not produce any CO_2 emissions or air pollution.

It is a highly efficient, low cost source of electricity.

Disadvantages of Hydroelectric Power

The newly created reservoir destroys habitat, which leads to a loss of biodiversity.

Sand and silt are not allowed to flow freely downstream. This results in beach erosion due to the lack of new deposition of sediment, and minimizes the deposition of natural fertilizing sediment downstream.

Dams require maintenance to function properly.

Hydrogen Fuel Cells

Hydrogen fuel cells have been widely anticipated because of their very high efficiency when compared to an internal combustion engine. In a hydrogen fuel cell, hydrogen's proton and electron are separated. The proton passes through a membrane, and ultimately combines with oxygen to form water. Water is the waste product of a fuel cell, which makes it an attractive alternative from a pollution standpoint. The separated electrons produce electricity which can be used to drive an electric motor or any other electrical device.

Advantages of Hydrogen Fuel Cells

Hydrogen fuel cells are very efficient.

There are no harmful emissions.

Hydrogen can be produced by splitting water and is relatively easy to store.

Disadvantages of Hydrogen as a Fuel Source

Fuel Cells are very expensive.

There is little to no hydrogen gas found in nature. Energy has to be used to create hydrogen by splitting water molecules. The most promising method of creating hydrogen would be to use wind or solar power to split the water molecules.

Hydrogen is a way to store energy, but not an energy source in itself.

There is no infrastructure in place to deliver hydrogen to the consumer.

Oikos

Oikos is the Greek root word of both ecology and economy. In the ancient Greek, the word means house or household. In this portion of the book we will deal with both the ecological aspects of environmental science and the role of man in the environment.

Ecology is the study of relationships formed between the **biotic** (living) and **abiotic** (non-living) factors found in nature. Environmental Science studies these relationships and interactions at several different levels. Environmental scientists also studies man's place in nature.

Populations

A **population** is a group of one species that live in the same area and can successfully interbreed. Several terms can be used when describing individuals within a population.

Generalists are organisms that have very broad niches. This gives them the freedom to have wide ranges and to fit into a variety of habitats. **Specialists** have a narrow niche and are very successful when conditions are favorable. However, if conditions are not favorable, specialists are more susceptible to extinction.

R-strategists are typically small organisms that have a high **biotic potential (r)** or population growth rate. They have large unprotected broods. These organisms generally have a short gestation time and individuals mature quickly. Insects and many plants are good examples of r-strategists.

K-strategists are generally larger and have fewer young. They invest large quantities of energy in rearing their young. Their population size does not vary nearly as much as r-strategists. Their population size is usually the maximum size (**carrying capacity [K]**) the environment will allow.

The size and density of a population is determined by the interplay of a specie's biotic potential and the environmental resistance. **Environmental resistance** includes the limited availability of shelter, food and water. Without environmental resistance a population would grow **exponentially**.

Logistic growth describes the growth of a population in response to environmental resistance.

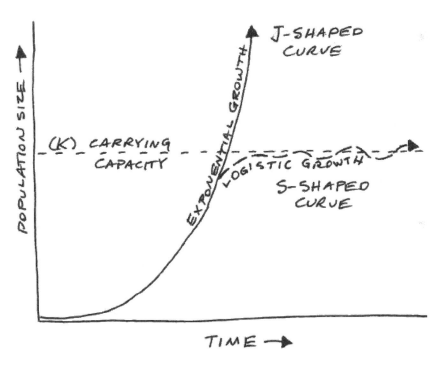

Population density describes the amount of a particular species in a given area. Many factors limit a population size; however, they all can be placed in two groups.

Density-dependent factors are environmental factors that have a greater effect on the more dense populations. An example of a density-dependent factor is the transmission of infectious disease.

Density-independent factors affect all populations regardless of size. Natural disaster, like fire and storms, are density-independent factors.

Individuals of a population are dispersed throughout the habitat in several ways.

Random dispersion occurs when a population is spread through a habitat by chance alone. An example of this would be the dispersion of dandelions on a hillside.

Uniform dispersion is the result of intraspecific competition and leaves the individuals of a population dispersed in an orderly fashion. Many seabirds exhibit uniform dispersion during the breeding season.

Clumped dispersion occurs when it is beneficial for the population to congregate together. This could be the result of isolated resources or for the safety of the species from predation.

Survivorship curves are used to illustrate reproductive strategies of species. Organisms have several different strategies to insure that their genetic information is passed on to the next generation. The following diagram illustrates these strategies.

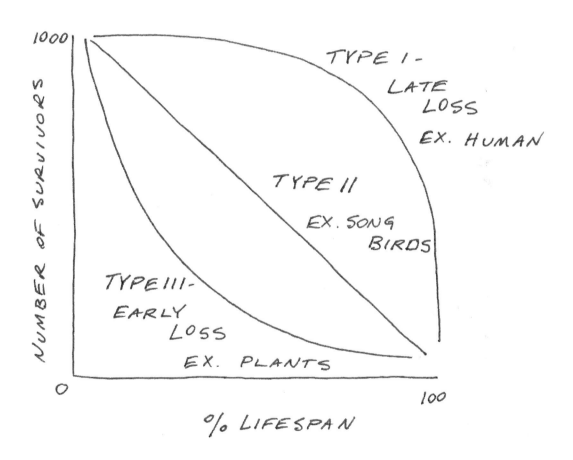

The Human Population

The current population is approximately 6.7 billion people! Throughout history, the human population has experienced exponential growth.

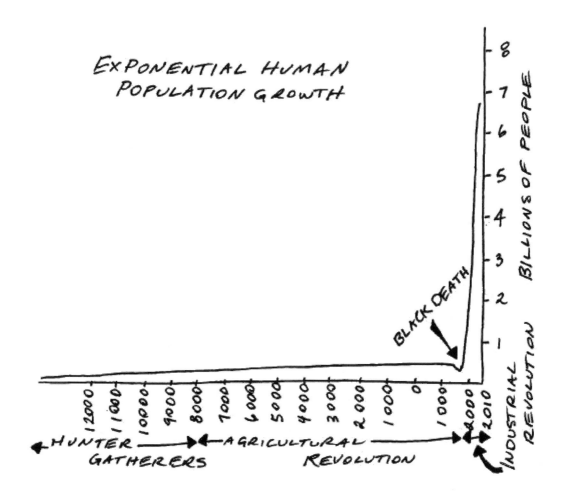

Several of the factors that have contributed to the exponential growth of the population include:

1. The Agricultural Revolution. The Agricultural Revolution led to food surpluses, and civilization, for the first time in human history. This stability led to an increase in population.

2. The Industrial Revolution and modern medicine. Improved living conditions and modern medicine have lead to longer life spans and decreased infant mortality.

These factors allowed death rates to decrease while birth rates remained high. This has led to rapid population growth.

There are several important terms to consider when discussing the human population. These terms include:

Total fertility rate: The average number of children per woman.

Replacement level fertility: The average number of children per woman necessary to sustain a population size. Ideally this would be two children per woman, one to replace the mother and one to replace the father. In reality, not all offspring survive to reproduce, so the average number or children per woman must remain slightly higher than two in order for a population to remain stable. Replacement level fertility can be only slightly higher than two in developed countries because of low infant mortality rates and good medical care. In developing countries where infant mortality rates are high and medical options are not as good replacement level fertility may be substantially higher than two children per woman.

Crude Birth Rates: The number of births per 1,000 people per year. Crude birth rate and total fertility rates are affected by education level and employment opportunities of women in a population. Cultural expectations, cost of child rearing, and availability of birth control also affect CBR and TFR.

Crude Death Rates: The number of deaths per 1,000 people per year. CBR is affected by working conditions, availability of good medical care, availability of good nutrition, access to clean water, and general living conditions.

Infant Mortality Rate: The number of infants per 1,000 births that die prior to their first birthday.

When birth rates exceed death rates, the size of a population will increase. Migration also plays a role in population size. **Emigration** refers to organisms leaving an area while **immigration** refers to individuals coming into an area.

Doubling time refers to the number of years that it takes for a population to double. The doubling time of a population can be estimated with the "rule of 70". Use the following equation.

Doubling time = 70/the percentage growth rate

For example, if a population is growing at 2% annually then:

Doubling time = 70/2% (keep as a whole number)

In this example, the number of years that would be required for this population to double is 35.

Age-Structure Diagrams (population pyramids) can be used to analyze the make-up of a population and project future growth trends.

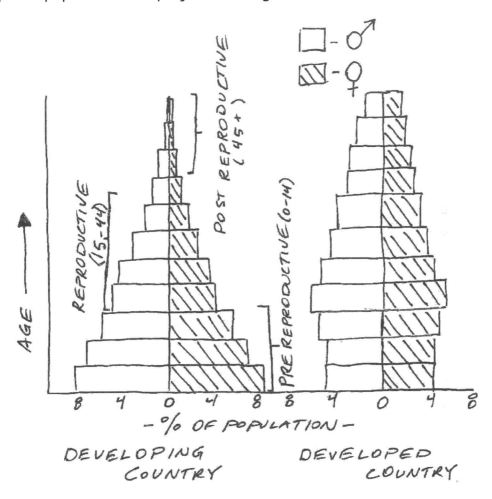

The **Demographic Transition** describes the transition of a country's population as it moves from a developing country to a developed country.

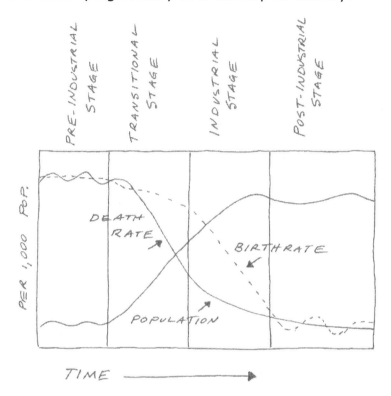

The **pre-industrial stage** is characterized by high birth rates and high death rates. This results in a stable population size.

During the **transitional stage** the death rate begins to drop with the advent of better food, medicine, and working conditions. Cultural traditions maintain a high birth rate and the population begins to grow rapidly.

During the **industrial stage** the birth rate begins to drop. The population continues to rise as the birth rate and death rate are still separated.

The **post-industrial** stage is characterized by the stabilization of population growth as the birth rates and death rates begin to come together again.

The question of whether or not the human population can, or should, be slowed down is often controversial. Because of our ability to innovate and problem solve it is very difficult to estimate what the carrying capacity of the human population is. The following steps could curb population growth:

1. Provide more educational and employment opportunities for women. Women in the workforce are much less likely to have large families.

2. Provide education about family planning options.

3. Some countries have initiated campaigns to educate its citizens about the benefits of smaller families. Others have initiated financial penalties on large families and financial incentives for small families.

Community Interactions

A **community** is all of the different populations of an area.

An **ecosystem** is the community and the non-living factors of an area.

The **biosphere** consists of the **lithosphere** (land), **hydrosphere** (water), and **atmosphere** (air) that contain life.

Life on earth is very diverse. Natural selection is the driving force behind this diversity. Understanding that organisms respond to their environment is critical for ecologists and environmental scientists. It allows them insight into ecology, environmental health, and agriculture problems which includes **pesticide resistance**.

The **niche** of a species is defined by its exploitation of resources and its interaction with the community. The **fundamental niche** is the full niche that a species can occupy. However, due to **interspecific competition** (between different species) and **intraspecific** competition (between same species), the **realized niche** is the part of the niche that the organism actually fills.

Resource partitioning occurs when species use resources at different times in different ways. This occurs in many ecosystems, but the tropical rainforests take it to the extreme.

In addition to competition, close interspecific relationships form in a community. These **symbiotic** relationships can be defined in four ways:

Parasitism is when one species benefits while harming the other (host). Ex. Tick and coyote

Mutualism occurs when both species benefit. Ex. Honey bee and flower

Commensalism is when one species benefits, but the other is neither helped nor harmed. Ex. Cattle egret and cow

Amensalism occurs when one species is harmed and the other is unaffected. Ex. Cattle creating trails in grass

Predator-prey relationships also form between members of a community. The populations of the species involved in this relationship are controlled in two ways. **Top-down control** is when the predator controls the population of the prey species. **Bottom-up control** states that the prey item directly controls the population of the predator. In reality, both have an effect on the size of each other's population.

In an ecosystem, all members of the community have a role in which they play. However, some species have a greater impact on an ecosystem than their biomass would suggest.   These species are known as **keystone species**. Examples of keystone species are the sea otters of the Pacific and their maintenance of the sea urchin population. Ecosystems are often defined by the presence of unique species.   These species are known as **indicator species**. Indicator species also provide insight into the health of an ecosystem.

The one way flow of energy from the sun maintains life on Earth.

Autotrophs or **producers** convert solar energy into chemical energy through photosynthesis. This energy then flows to **heterotrophs** when they consume the plants. Organisms that eat only producers are known as **primary consumers** or **herbivores**.   **Secondary** and **tertiary consumers** obtain their energy from consuming other heterotrophs.   Organisms that only eat heterotrophs are **carnivores**. Organisms that eat both producers and consumers are **omnivores**. **Detritivores** consume dead organisms. **Decomposers** further breakdown the **detritus** and convert it into inorganic nutrients that plants can readily use.

The feeding relationships of an ecosystem are often presented in **food chains** and **food webs.**

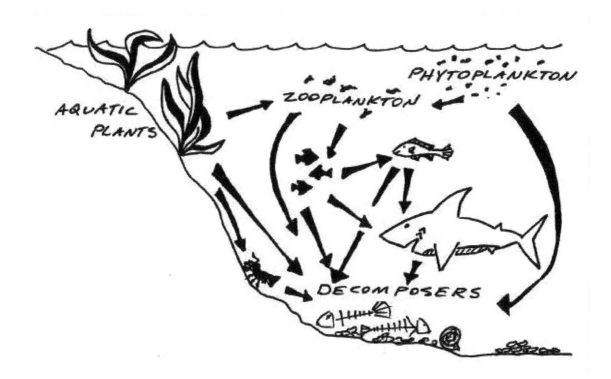

The transfer of energy in an ecosystem can also be illustrated with **ecological pyramids**. These ecological pyramids can be used to represent energy efficiency, biomass, and numbers in each feeding level or **trophic** level. All of these ecological pyramids point out the fact that energy, biomass and numbers (generally) decrease in higher trophic levels.

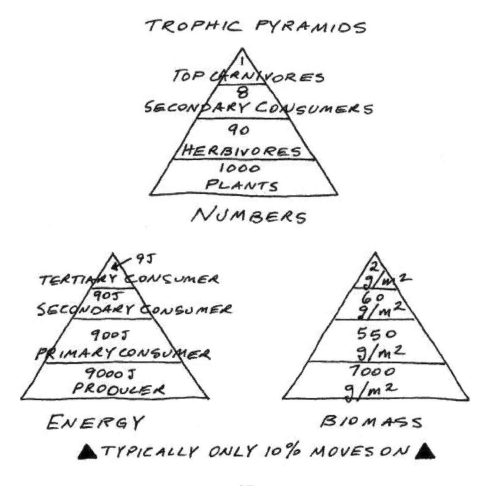

TROPHIC PYRAMIDS

1
TOP CARNIVORES
8
SECONDARY CONSUMERS
90
HERBIVORES
1000
PLANTS

NUMBERS

9J
TERTIARY CONSUMER
90J
SECONDARY CONSUMER
900J
PRIMARY CONSUMER
9000 J
PRODUCER

ENERGY

2
g/m^2
60
g/m^2
550
g/m^2
7000
g/m^2

BIOMASS

▲ TYPICALLY ONLY 10% MOVES ON ▲

When new communities are established or existing communities are disturbed, a process called **ecological succession** will take place. This process describes the steps in which the new communities are formed. Each step is made by an organism changing its surroundings. These changes allow subsequent species to survive and further alter the environment.

Primary succession is the establishment of a community in a soil-less environment. This could be in an area of a lava flow or the retreat of a glacier. **Pioneer species** like mosses and lichens (symbiotic relationships of fungi and algae) are the first inhabitants. As they grow, they release acids that break down rock into smaller sediments. They further add to the creation of soil when they die and decompose. Small plants and insects arrive next, providing even more nutrients and mass to the soil. Over time, if precipitation is sufficient, larger plants and trees will become established.

Secondary succession occurs in ecosystems that have been disturbed by fire or other natural disasters. It also takes place in land disturbed by man that is allowed to return to its native state.

Biogeochemical Cycles

The Carbon Cycle

Photosynthesis and cellular respiration are key activities in the biogeochemical cycle of carbon. During photosynthesis, plants take inorganic carbon dioxide and convert it into glucose, a high-energy sugar molecule. Consumers and producers use glucose in respiration and return carbon back in the environment as carbon dioxide. Decomposition of detritus also releases carbon dioxide into the atmosphere. Fossil fuels were formed when incomplete decomposition was accompanied by heat, pressure, and burial. Fossil fuels include oil, coal, and natural gas. When man burns these fossil fuels, carbon dioxide is rapidly released back into the atmosphere. Since carbon dioxide is a greenhouse gas, this practice has led many scientists to believe drastic climate change is imminent. Removal of large tracts of forests also adds to the release of carbon dioxide. The ocean also acts as a sink or reservoir for carbon dioxide.

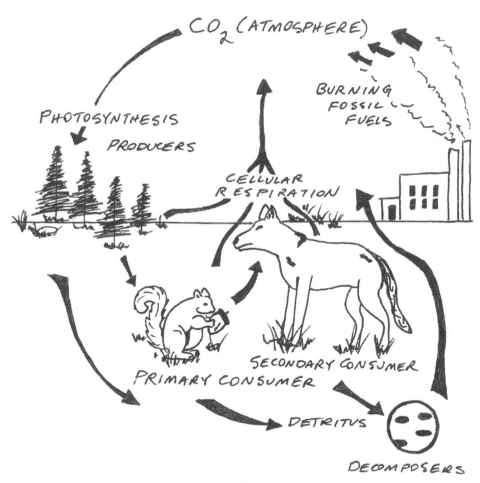

Nitrogen Cycle

Nitrogen makes up seventy-eight percent of the atmosphere. In order for it to be used by organisms it first has to be converted into a useable form.   The following are the major steps in the nitrogen cycle:

Nitrogen fixation is the conversion of atmospheric nitrogen gas (N_2) in the atmosphere into ammonia (NH_3) or ammonium (NH_4). Lightening and specialized bacteria found in the soil accomplishes this step.   **Rhizobium** bacteria, which form a mutualistic relationship with legumes, are examples of nitrogen-fixing bacteria.

Nitrification occurs when ammonium (NH_4) is converted into nitrates.

When detritus is decomposed, ammonia is released in the process of **ammonification**.

Assimilation occurs when plants incorporate the usable forms of nitrogen into organic molecules, which ultimately are incorporated into consumers.

Denitrifying bacteria converts nitrates into atmospheric nitrogen in the **denitrification process**.

Fertilizers from agricultural runoff disrupt the nitrogen cycle. This can also lead to **cultural eutrophication** of water sources.

The Nitrogen Cycle

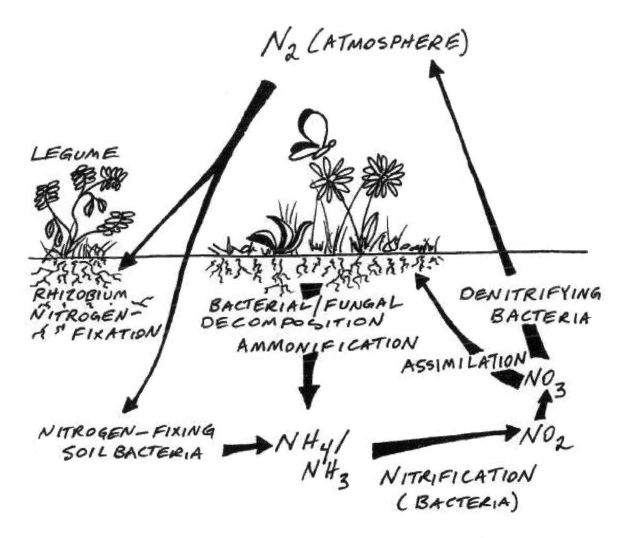

N_2 (ATMOSPHERE)

LEGUME

RHIZOBIUM
NITROGEN—
1st FIXATION

BACTERIAL/FUNGAL
DECOMPOSITION
AMMONIFICATION

DENITRIFYING
BACTERIA

ASSIMILATION

NO_3

NITROGEN—FIXING
SOIL BACTERIA → NH_4/
NH_3

NO_2

NITRIFICATION
(BACTERIA)

Phosphorus Cycle

The phosphorus cycle is a **sedimentary** cycle. Unlike the other major biogeochemical cycles it does not have an atmospheric component. Therefore, the phosphorus cycle is a slow process which often causes it to be a **limiting factor** in ecosystems. A limiting factor determines the growth of a population. Phosphorus is released from rocks by weathering. **Immobilization** occurs when phosphorus is converted into an organic form and not available for assimilation by plants. In **mineralization**, phosphorus is in an inorganic form, readily absorbed by plants. Much like the nitrogen cycle, bacteria play an important role in the phosphorus cycle. Phosphorus is often a key component of fertilizer.

Sewage and Solid Waste

THERE IS NO AWAY! With that in mind, let's take a look at what happens to your waste when you throw away your garbage or flush your toilet.

Wastewater Treatment

Every day millions of Americans flush their toilet, and don't give a second thought to the final destination of the toilets contents. While it might be nice to imagine the waste completely disappearing once it is out of site, in reality treating wastewater is a complicated process with a solid waste problem waiting at the end of it. Wastewater travels from the toilet, through sewer lines, and ultimately ends up at a wastewater treatment facility. The wastewater treatment process can usually be broken down into primary treatment, secondary treatment, and tertiary treatment.

In **primary treatment** (a physical process) large objects (use your imagination) are removed from the wastewater with large screens. As the water flows through the screens, the objects are filtered out. The resulting solid garbage must be disposed of in a landfill. The water is then diverted to settling tanks where remaining particulate matter is allowed to settle out of the water. The water is then sent to secondary treatment.

In **secondary treatment** (a biological process) the wastewater is mixed with oxygenated water. This encourages the bacterial breakdown of the organic matter left in the wastewater. The bacteria and particulate matter settle out of the water creating sewage sludge. The sludge is removed, dried, and then either disposed of in landfills, incinerated, or used for fertilizer. The water is then disinfected using chlorine, ozone, or UV light. In some cities, the water is then put through tertiary treatment.

Tertiary treatment involves using physical or chemical methods to remove inorganic nutrients such as nitrogen or phosphorus from the water. Removing these chemicals will help combat eutrophication in areas where the water is discharged.

In rural areas where it doesn't make sense to build a wastewater treatment plant, residents rely on septic systems which serve as individual wastewater treatment plants. Sewage from a home is held in an underground tank where it is broken down by bacteria. The leftover water is allowed to leach into the soil through underground pipes. In newer systems, the water is disinfected and used for lawn irrigation.

In many areas, city planners are looking to nature for innovation in wastewater treatment. There have been several facilities built that pass sewage through artificial wetlands. This allows gravity and biological processes to treat the sewage, with little to no negative impact.

Sewage Treatment

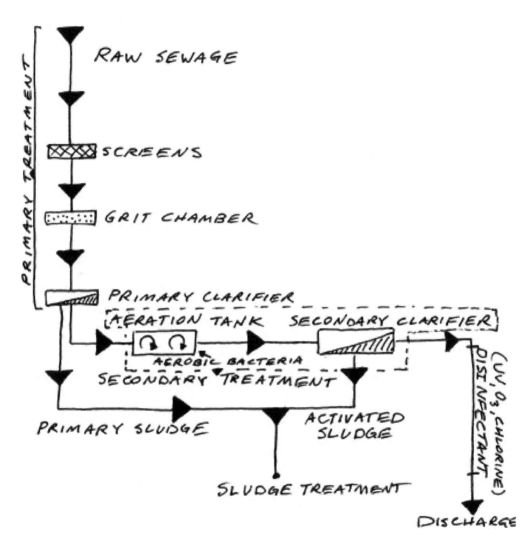

Landfills

Now that we have dealt with your toilet, let's take a look at what happens to the trash that you throw out each day. After pickup by the garbage man, the fate of your garbage most likely lies in a landfill or incinerator.

Most of the municipal waste in the United States ends up in landfills. The simplest form of landfilling involves collecting and dumping trash in open piles. The piles are then left to decompose or sometimes burned to make space. This was practiced for a long time in the United States. The problem with a giant pile of trash is that any of the pollutants in the trash are free to runoff. The trash heap is also a prime habitat for rodents and other disease vectors. In the U.S., this method has been predominantly replaced with sanitary landfills. **Sanitary landfills** involve burying trash in a pit that has a clay bottom and sides and a plastic liner. The trash is layered with soil. The soil helps the trash decompose faster, and helps keep animals out of the trash. When the landfill is full, it is covered with plastic, clay, and topsoil. The retired landfill can then be used for further development. There have been many city parks built on top of old landfills.

When rain soaks through a landfill, the resulting contaminated liquids are called **leachate.** Sanitary landfills have a leachate collection system that runs beneath the landfill, and groundwater is continuously monitored to make sure that the leachate is not seeping into the groundwater.

Sanitary Landfill

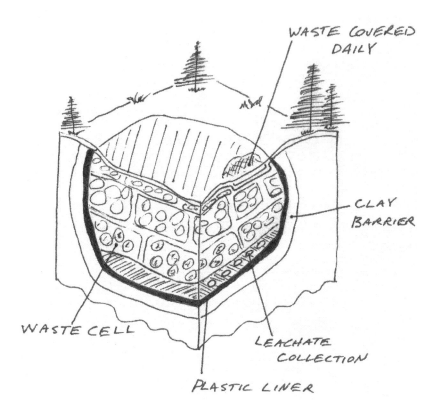

WASTE COVERED DAILY

CLAY BARRIER

WASTE CELL

LEACHATE COLLECTION

PLASTIC LINER

Incineration

In some areas, waste is incinerated to reduce the burden on landfills. When garbage enters an incineration facility, the metals are usually removed, and the garbage is shredded. The garbage is then burned at very high temperatures. The left over liquid and ash must be treated and buried in a landfill. The ash can contain many toxic chemicals such as heavy metals or dioxin. The emissions from the incineration plant must pass through a scrubber to remove air pollutants such as sulfur dioxide. Incineration drastically reduces the volume of trash that is placed into landfills. In waste-to-energy incinerators, the heat that is generated by burning trash is used to create steam. The steam is then used to spin a turbine attached to a generator. Electricity is then produced in the same way as most other power plants.

Reduce, Reuse, Recycle

"Reduce, Reuse, Recycle..." isn't just a catchy line from a Jack Johnson song. The "three R's" play an important role in reducing the total volume of garbage. Our first priority in waste reduction should be to reduce. **Reducing** involves purchasing goods that have less packaging material, repairing instead of replacing, and using products for their full lives instead of making unnecessary upgrades. Yes, this is directed to those of you who buy a new cell phone every six months!

Reusing involves simple actions like reusing plastic bags, cardboard boxes, avoiding disposable items, etc... It could also mean buying a used car instead of that new one that you have had your eye on. Both reducing and reusing result in less waste entering the waste stream.

After we have cut down our waste generation by reducing and reusing it is important to **recycle** whatever we can prior to sending garbage to a landfill. Recycling generally involve a local facility where citizens can drop off metals, plastics, and paper. Many cities have curbside recycling where garbage trucks pick up the recyclables and take them to a **Materials Recovery Facility.** Here the materials are sorted and sent to manufacturers who use the waste as raw materials for new products. Recycling must be economically feasible for cities to continue to invest in recycling programs. It is therefore important to support companies who offer products made of recycled materials.

The Fine Print

The **Pollution Prevention Act** focuses on the prevention of pollution through reduction of waste, pollution, and inefficiency at the source.

The **Resource Conservation and Recovery Act (RCRA)** allows the EPA to regulate the handling, storage, and disposal of hazardous waste.

The **Comprehensive Environmental Response, Compensation, and Liability Act (CERCLA)** created a federal "superfund" that would be used to clean up abandoned hazardous waste, or emergency discharges of hazardous waste. This is often called the **Superfund Act**.

The **Toxic Substances Control Act** allows the EPA to keep track of industrial chemicals produced or imported into the United States.

Environmental History

Environmental history is rarely tested directly on the exam. However, there are some important people, dates, etc. that you should know about. It is also important to have some historical context for understanding the social and political aspects of environmental science.

Important People

Theodore Roosevelt: U.S. President from 1901-1909. Roosevelt is known as a conservationist who protected vast amounts of land during his presidency.

Aldo Leopold: Author of "A Sand County Almanac". Leopold's work helped to inspire the modern environmental movement.

John Muir: Founded the Sierra Club in 1892. John Muir argued for the creation of national parks.

Rachel Carson: In 1962 published <u>Silent Spring</u>. The book brought attention to problems associated with the use of pesticides.

Richard Nixon: US President. Established the Environmental Protection Agency in 1970.

Jimmy Carter: US President from 1977-1981. Carter Urged the creation of the Department of Energy, and the formation of the "Superfund Act".

Important Events and Eras

Hunter-Gatherers: Prior to the agricultural revolution, early humans existed as hunter-gatherers. They survived by hunting, scavenging, and collecting wild plants. Early humans were highly mobile, and the human population was very small. The environmental impact of the hunter-gatherers was relatively low due to the small population size. Hunter-gathers were not without environmental impact entirely. Early humans were in many instances responsible for the extinction of many of the megafauna on the planet.

Agricultural Revolution: The first Agricultural Revolution began as many as 10,000 years ago. During this period of time, civilization made the shift from a hunter-gatherer society to an agricultural society. We collectively learned how to cultivate plants which led to more reliable food production. Since the industrial revolution, agriculture has expanded exponentially with the development of farm machinery, pesticides, herbicides, fertilizers, and advanced irrigation techniques.

Industrial Revolution: The Industrial Revolution began in the mid 1800s in Europe and quickly spread to the United States. The development of the steam engine, internal combustion engine, and the use of fossil fuels brought about the Industrial Revolution. During this period, industry and agriculture grew significantly as mass production and new technological advances were developed. The mining and use of fossil fuels increased as well as the production and use of chemicals.

Globalization: The world is currently in a period of globalization. This is characterized by international commerce, increased travel, sharing of cultures and traditions, as well as an increase in the middle class. Globalization has been seen as an unprecedented opportunity for the sharing of ideas and resources, but it has also resulted in a loss of cultural diversity and biodiversity.

Even More Fine Print

The **National Environmental Protection Act (NEPA) (NEPA)** requires that all branches of the government give due consideration to environmental concerns when federal activities are carried out.

The **Emergency Planning and Community Right to Know Act** requires that annual reports be made regarding the release of toxic chemicals into the environment.

The Extras

Words You Need to Know

From time to time, there will be words on the APES exam that may throw you for a loop. Here is a list of words that don't necessarily have anything to do with environmental science, but have shown up on the exam in the past. These are all words that our students have had a hard time with in the past, and it is necessary to have a good vocabulary in order to understand what the exam questions are asking you.

Abundant: Abundant simply means plenty, or a lot. Coal reserves in the United States are relatively abundant.

Anthropocentric: Human centered. The decision to build the new shopping mall on the site of the town's nature preserve was anthropocentric.

Anthropogenic: Human caused. There are many anthropogenic sources of air pollution.

Causative, causal, causation: Three different words that all imply that one thing "causes" another thing to happen. Agricultural runoff is a causative agent of cultural eutrophication.

Collaboration: Working together. Many governments collaborated to draft and enact the Kyoto Protocol.

Combustion: Having to do with fire. The incomplete combustion of fossil fuels contributes to acid rain.

Commercial: Having to do with business, or economy. A commercial fisherman is one who fishes to make a profit as opposed to a recreational fisherman who fishes for fun.

Compensation: Payment. The company received compensation from the government to store the nuclear waste.

Consumption: Use. Consumption of seafood has increased steadily since 1950.

Contention: A point of argument. The farmers main point of contention was that the new pesticide legislation would result in decreased profits.

Correlation: Relationship. There is a positive correlation between agricultural runoff and cultural eutrophication.

Devoid: Without. My environmental science teacher must be devoid of a soul!

Efficiency: Efficiency refers to the amount of work, or amount of product, that is received for a given amount of energy that is expended. This term is likely to be used in questions about different types of energy or fuels. Incandescent light bulbs operate at approximately 5% efficiency.

Emigration: Leaving a country or area. The local famine forced many families to emigrate to another area.

Employs: Uses. Grassroots environmental organizations employ volunteers to support their cause.

Equatorial: Pertaining to the equator. Equatorial regions receive more sunlight annually than the polar regions of the earth.

Fluctuation: Change. The salmon population in the pacific northwest fluctuates regularly.

Fragmentation: Separation. Habitat fragmentation disrupts the ability of certain organisms to migrate.

Immigration: Moving into a country or area. The population increase in the United States is due almost entirely to immigration.

Implicated: Implicated means that someone or something has been shown to be associated with an event. DDT was implicated in the disruption of the reproductive process in Brown Pelicans.

Inadvertent: Unintended, or by accident. An inadvertent consequence of the local feedlot was the runoff of animal waste into the local lake.

Incidence: Occurrence or happening. The incidence of heart disease in the United States is relatively high.

Insoluble: Not soluble, won't dissolve. Fat soluble vitamins are insoluble in water.

Mean: Mathematic average. The mean income for Americans is $50,000.

Municipal: City.   Municipal solid waste is the primary material in the local landfill.

Proliferation: Expansion.   The proliferation of nuclear weapons is of great concern to the United States.

Proportional: Relationship of parts to each other, or to the whole.   The EPAs response to the companies' environmental problems was proportional.

Uniformity: Uniformity implies sameness.   The uniformity of the fast food restaurant's product was remarkable.

Viable: Capable of sustaining life, or likely to succeed.   The government struggled to pass viable legislation on climate change.

Graphing and Math

On the free response portion of the exam you may be asked to draw a graph. If this happens you should be very happy. These may be the easiest points on the exam, but unfortunately they are often missed when students are not careful. If you draw a graph make sure that it has a descriptive title, clearly label your axes with what you are measuring and how you are measuring it, and make sure that you plot your graph appropriately with the independent variable (x axis) and the dependent variable (y axis) in the correct place.

If you are asked to calculate something you must make sure to **show your work** no matter how simple it seems. Also make sure to **label your answers with the appropriate units**.

You will often have to use **scientific notation** to solve problems on the free response portion of the exam. Brushing up on scientific notation will save you a great deal of time when working with large numbers.

Dimensional analysis is a very handy tool to have for questions that require you to do energy conversions and comparisons, such as comparing the output of a coal power plant to that of a wind farm.

You will often find word problems in the multiple choice portion of the exam. These will most likely only require simple division and multiplication. If word problems are not your forte, then you may want to practice a few before the exam

You are not allowed to use a calculator on the APES exam, so you should practice working with fractions, multiplication, percentages, and division by hand. With no calculator allowed, the test questions must be written so that you can complete them in a reasonable amount of time. Most of the calculations will involve multiples of 10, and only require one, two, or maybe three steps. **If you begin to work on a problem and it seems like it is going to be extremely complicated, then you are most likely looking at the problem wrong.** Mentally regroup, reread the question, and give it another try.

Designing an experiment

From time to time, you may be asked to design an experiment on the free response portion of the exam. Here are some important things to include in you answer.

1. Define the experimental and control groups. Only give one independent variable.

2. Always say that you would repeat your experiment.

3. Mention having a large sample size. The more samples you take, the greater the validity of your results.

4. Make sure to mention that you would control all variables (the constants) other than the independent variable that you intend to manipulate.

5. Remember that you are not actually conducting the experiment, it just needs to be feasible and scientifically sound.

When All Else Fails...

When times get tough, and you just don't know the answer to the question you may want to consider the following options. These are standby answers that work for almost any question about the impacts of an environmental problem.

Option #1: The incomplete combustion of fossil fuels. The incomplete combustion of fossil fuels results in the production of CO_2, sulfur compounds, nitrogen compounds, and contributes to photochemical smog. This could be used as a contributor to acid raid, contributor to global warming, contributor to air pollution, etc.

Option #2: Loss of habitat leading to the loss of biodiversity. This one is pretty simple. Almost any environmental problem involving humans can be related to a loss of habitat which also leads to a loss of biodiversity. Housing development, monoculture, development of infrastructure, air pollution, water pollution, global warming, oil spills, energy production, food production, water diversion products, and deforestation can all lead to loss of habitat, which in turn endangers the species that live in that habitat.

Benjy Wood lives in Plano, Tx with his wife and three children. He is a high school science teacher with a long track record of success teaching AP Environmental Science. Benjy is a former President of the Texas Association of Biology Teachers, and has served as an exam reader and a science education consultant specializing in environmental science and AP Exam preparation. Benjy and Dave Co-founded BioGuild Consulting in 2008.

Dave Holbert lives on a small farm in Clay County with his wife and children. He is a high school teacher with a short track record of success teaching AP Environmental Science. Dave serves as a mentor teacher for the Regional Science Collaborative. He is also the recipient of the 2009 Shell Distinguished Teaching Award and the 2009 West Teaching Excellence Award. Dave also serves as a science education consultant specializing in pre-AP middle school and biology training and environmental science AP Exam preparation.

This book is definitely a work in progress. We truly welcome your comments, so that we may better serve future generations of APES students. We welcome your comments at:

www.bioguildconsulting.com

Made in the USA
San Bernardino, CA
15 December 2016